T0199652

Spark Erosion Machining: MEMS to Aerospace

Edited by

Neelesh Kumar Jain

Professor, Department of Mechanical Engineering
Indian Institute of Technology, Indore, India

Kapil Gupta

Associate Professor, Mechanical and Industrial
Engineering Technology University of Johannesburg,
Johannesburg, South Africa

CRC Press
Taylor & Francis Group
Boca Raton London New York

CRC Press is an imprint of the
Taylor & Francis Group, an **informa** business

A SCIENCE PUBLISHERS BOOK

CRC Press
Taylor & Francis Group
6000 Broken Sound Parkway NW, Suite 300
Boca Raton, FL 33487-2742

Version Date: 20210502

International Standard Book Number-13: 978-1-498-78793-2 (Hardback)

Visit the Taylor & Francis Web site at
http://www.taylorandfrancis.com

and the CRC Press Web site at
http://www.crcpress.com

Preface

The more stringent requirements for shape and size of engineered products made from a wide range of engineering materials have largely been responsible for the development of advanced or unconventional or non-traditional machining processes and subsequently replacement of the conventional machining processes by them. Spark erosion machining (SEM), most commonly known as electrio discharge machining (EDM), is one of the advanced machining processes that has been established as a prominent process in micro-electro mechanical systems (MEMS) to aerospace industry to fulfil the specialized requirements of drilling, profile-cutting, slitting, machining of 3D complicated shapes, etc. Whether machining of micro-parts, typical shapes or features, or cutting of difficult-to-machine materials, SEM, its variants such as wire-SEM, spark erosion drilling (SED), wire spark erosion grinding (WSEG), and the hybrid processes based on it i.e. electrochemical spark machining (ECSM), travelling wire-ECSM (TW-ECSM), etc. have been found very successful from the point of view of quality, productivity, and sustainability.

This book provides a basic and advanced level knowledge on different aspects and applications of SEM process and the processes based on it in nine chapters. It starts with an introductory Chapter 1 on SEM that introduces the process and discusses all important aspects such as basic principle, mechanism, and variants of this process. Chapter 2 discusses different aspects of machining of shape memory materials by WSEM process. Issues related to precise manufacturing of the biomedical components by SEM based processes are discussed in Chapter 3. Comprehensive information on machining of aerospace materials by SEM process is provided in Chapter 4. WSEG process is discussed in Chapter 5. Important aspects of different hybrid processes based on SEM are described in Chapter 6. Vibroacoustic diagnostics of SE-based processes are presented in Chapter 7. Whereas, Chapter 8 recommends interventions and ways to improve sustainability of SE-based processes. The book concludes with Chapter 9 describing different techniques used to optimize parameters of SE- based processes with an objective to enhance their performance.

The main purpose to bring-out this book is to facilitate the academic and research community by offering the theoretical background, novel aspects, research advances and applications of SEM process and its variants. This

book also intends to enable and encourage the researchers to explore the
field and develop this process further with an objective to find the solutions
for the major industrial problems and to work for the societal and human
benefits.

    We sincerely acknowledge our heart-felt gratitude towards the
contributing authors for their time and efforts and CRC Press of Taylor &
Francis Group LLC for this opportunity and their professional support. We
hope this book will be valuable addition to enhance knowledge and research
in the field of SEM process.

April 2020

Neelesh Kumar Jain
Kapil Gupta

Contents

PART I
Introduction

Spark Erosion Machining

Andrew Rees

College of Engineering, Swansea University Bay Campus, Crymlyn Burrows,
Swansea, SA1 8EN, United Kingdom
Email: andrew.rees@swansea.ac.uk

1.1 Introduction

Spark Erosion Machining (SEM), more commonly known as electric
discharge machining (EDM), belongs to the category of *thermal-type
unconventional* or *non-traditional* or *advanced* machining processes. It
has been replacing drilling, milling, grinding and other traditional
machining processes at a fast pace, in various industrial applications for
the manufacture of components and products from difficult-to-machine
materials. Over the last 50 years, SEM has transformed into one of the
most advanced machining processes in the market. This transformation
has been possible with the integration of computer numerical control
(CNC) and state-of-the-art power supply technology (Bleys and Kruth
2001). Consequently, even 6-axis CNC SEM machines are now available
and can produce very complicated components having mirror finish
surfaces, with high accuracy and precision, from a wide range of materials
that are otherwise impossible or difficult to machine by traditional machining
processes (Ramaawmy et al. 2005).

The concept of SEM can be traced back to 1770, when Joseph Priestly, an
English chemist, discovered the erosive property of electrical sparks (Webzell
2001); but this could not be utilized for material removal purposes until 1943,
when two Soviet scientists, Boris Lazarenko and Natalia Lazarenko, through
their ground-breaking research, generated controlled sparks between two
electrodes of opposite polarities, using a resistance-capacitance (RC) circuit.
Subsequently, with its growing merits and capabilities, the SEM process has
been intensely sought by the manufacturing industry. Adoption of advanced
servo-control has also helped in optimizing the gap between the tool and the
workpiece, making material removal more efficient.

With SEM, the workpiece material is removed by the erosive action by a
controlled series of electrical discharges occurring between the tool electrode

and the workpiece electrode, both immersed in a dielectric fluid and separated by a small working gap. It includes ignition, melting and ejection phases (Figures 1.1, 1.2 and 1.3). The *ignition* phase (Fig. 1.1) is the preparatory phase for the discharge of a spark. It is initiated as soon as the pulsed DC voltage with predetermined pulse frequency is applied between the tool electrode and the workpiece, by a pulsed power source, causing the electric field in the inter-electrode gap (IEG) to exceed its critical value. This causes the cathode to emit a stream of electrons which then accelerates towards the anode. This streamer acts as a precursor for the effective breakdown (i.e. ionization) of the dielectric as, during their movement towards the anode, these electrons collide with the molecules of the dielectric and, when the applied voltage attains a sufficient value (more than the breakdown voltage of the dielectric), the dielectric molecules break down (become conducting) through ionization, hence releasing a stream of secondary electrons. The avalanche of such electrons is seen as a spark discharged during the pulse-on time at the location with minimum IEG. The spark discharge creates a high-pressure and high-temperature plasma channel consisting of free electrons and positively charged ions surrounded by a gaseous mantle. The electrons accelerate towards the anode and heat it up after striking it, whereas the positively ions strike and heat up the cathode. Concentration of high-power plasma channels on both the electrode and the workpiece causes them to melt and evaporate during the *melting* phase (Fig. 1.2). Relatively more material is removed from the anode than the cathode due to three reasons: (i) the momentum with which electrons strike the anode being much more than that of the momentum with which ions strike the cathode, (ii) creation of a thin carbon film on the cathode (due to pyrolysis of the dielectric fluid), and

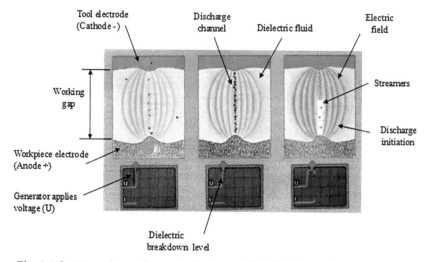

Fig. 1.1: Ignition phase of a spark discharge (OEL-HELD GmbH publication)

(iii) development of compressive force on the cathode. Most of the molten material is flushed out of the IEG by a continuous high-velocity and high-pressure supply of dielectric fluid during the pulse-off time, thus creating craters on both the electrode and the workpiece, while the remaining molten materials re-solidify in the *ejection* phase (Fig. 1.3). The dielectric also deionizes, forming its molecules again during the pulse-off time. This cycle of events continues, with each discharge of spark melting and evaporating a very small amount of material from both the tool and the workpiece. Since,

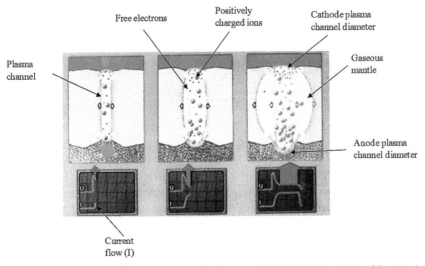

Fig. 1.2: Melting phase of a spark discharge (OEL-HELD GmbH publication)

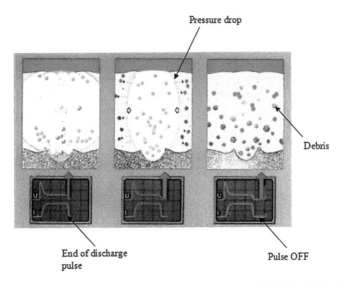

Fig. 1.3: Ejection phase of a spark discharge (OEL-HELD GmbH publication)

the location of minimum IEG keeps on changing after the occurrence of each spark, the spark keeps on travelling throughout the IEG. Longer pulse-on times increase the amount of materials melted and removed. The polarities of the workpiece and the tool are chosen in such a way that more material is removed from the workpiece than that from the tool. This can be achieved by an appropriate selection of electrode materials, dielectric and SEM process parameters (Toren et al. 1975, Meeusen 2003, Kunieda et al. 2005, Rajurkar et al. 2006, Abbas et al. 2007, Yeo et al. 2008).

1.2 Types of SEM Processes

Various derived and hybrid SEM processes have been developed over the past six decades. Initially, SEM was mainly used for manufacturing different types of dies and molds; therefore, SEM is also referred to as die-sinking SEM. Of the numerous derived SEM processes, wire spark erosion machining (WSEM) is the most important.

1.2.1 Die-sinking SEM

The die-sinking SEM process (Fig. 1.4) is the oldest version of SEM widely used for near net-shape manufacturing of dies and molds for aerospace, automotive and biomedical applications (Kalpajian and Schmid 2003). Copper, tungsten or tungsten carbide can be used to manufacture tool electrodes, giving a shape complimentary to the desired shape to be produced on the die or mold required. The tool electrode travels towards the workpiece sunk in a suitable dielectric until the desired IEG is achieved. Subsequently, workpiece material is removed through the erosive action of a series of discharges of short-duration sparks between the electrode and

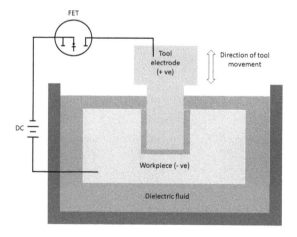

Fig. 1.4: Schematic of die-sinking SEM process

the workpiece (Livshits 1960, Jameson 2001). As there is no direct contact between the tool and the workpiece die, no mechanical stress on the die or mold, or cutter chatter, or vibration is generated (Kalpajian and Schmid 2003). The die-sinking SEM process is capable of machining any electrically conductive die material, regardless of its hardness (Houman 1983). It can be used for the following types of die materials:

1.2.1.1 Die-Sinking SEM of Heat-treated Materials

Die-sinking SEM has replaced conventional machining processes (such as milling) for heat-treated materials in the manufacture of dies and molds as its performance is independent of material hardness, whereas milling can machine materials with hardness in the range of 30-35 HRC only (Bayramoglu and Duffill 1994). Moreover, SEM does not induce any mechanical stress on the die material (Arthur et al. 1996).

1.2.1.2 Die-Sinking SEM of Ceramics

Mohri et al. (1996) demonstrated that using an auxiliary or assisting electrode can help in the machining of ceramics through any SEM process. They successfully diffused conductive particles from an assisting electrode to the surface of sialon ceramics or silicon nitride (Si_3N_4). Konig et al. (1998) characterized ceramics into non-conductors, natural conductors and conductors, and demonstrated that SEM of non-conducting ceramics can be facilitated by doping them with conductive elements. Lee et al. (1998) proved that a hybrid process enhances the circulation of the dielectric in the IEG and reduces the thickness of the white or recast layer. Sanchez et al. (2001) demonstrated the feasibility of SEM of boron carbide (B_4C) and silicon-infiltrated silicon carbide (SiSiC) by combining ultrasonic vibrations with SEM. All these developments have turned SEM and SEM-based processes into competitive technologies for machining advanced ceramics.

1.2.1.3 Die-Sinking SEM of Composite Materials

Performance characteristics such as material removal rate (MRR), electrode wear and the resulting surface quality and integrity are crucial in die-sinking SEM to expand its applications to composite materials. Yan et al. (2000) examined the machining response of metal matrix composite (having 6061 Aluminum alloy reinforced with Al_2O_3) to rotary SEM coupled with a disk electrode to expand the applications of SEM. Muller and Monaghan (2000) compared SEM with other non-conventional machining technologies such as laser beam machining and abrasive water jet cutting. They concluded that though the SEM of reinforced metal matrix composites is slow, it results in relatively lower sub-surface damage.

1.2.2 Wire-Spark Erosion Machining (WSEM)

Wire spark erosion machining (WSEM) is a process derived from SEM, which uses a continuously travelling thin wire as the negatively charged tool electrode to machine a positively charged workpiece through a succession of discharges occurring between them. The wire can be made of copper, brass, tungsten or molybdenum, with a diameter in the range of 100-250 μm. It is fed under constant tension at a predetermined feed rate (Fig. 1.5). Since the size of the wire is very small (which makes it susceptible to breakage by spark discharges) as compared to the pre-shaped tool electrode used in SEM, WSEM uses deionised or distilled water as the dielectric, with a dielectric strength substantially lower than that of hydrocarbon oils used for regular SEM. Furthermore, the pulse frequency herein is in the order of MHz (kHz in SEM) and the applied voltage is low. Also, continuous feeding of wire avoids higher wear rates and adverse effects on the quality of the workpiece (Huntress 1978).

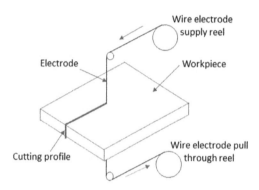

Fig. 1.5: Working principle of WSEM

The biggest advantage of the WSEM process is the elimination of elaborate design and manufacturing of pre-shaped electrodes as required in SEM and die-sinking SEM. During the WSEM operation, the wire makes several passes along a pre-defined profile to obtain the desired dimensional and surface finish requirements. It can produce surface roughness down to Ra 0.04. Typical WSEM cutting rates are 300 mm²/min for a 50 mm thick D2 tool steel workpiece and 750 mm²/min for a 150 mm thick aluminium workpiece (Kalpajian and Schmid 2003).

1.2.2.1 WSEM Applications

WSEM has proven to be a techno-commercial solution for meeting the extensive demands of automotive, aerospace, mold, tool-making and die-making industries. It is used to cut intricate profiles with high precision and to machine electrically conductive materials into any shape, regardless of their hardness, strength or toughness (Ho et al. 2004). WSEM has also found uses in biomedical and dental fields, and microelectronics and computer peripherals (Stovicek 1993).

Many researchers have investigated the applications of WSEM in the machining of modern composites and advanced ceramic materials. Levy and Wertheim (1988) and Luo et al. (1992) studied the machining of silicon wafers and sintered carbide dies. Kruusing et al. (1999) compared laser cutting with WSEM for the machining of NdFeB and 'soft' MnZn ferrite magnetic materials. Rhoney et al. (2002) used WSEM for the dressing of a rotating metal-bonded diamond wheel.

Industries have been focussing on unattended machining by WSEM, for which CNC machines play a significant role by controlling machining strategies to prevent frequent wire breakages, hence facilitating automatic wire threading (Ho et al. 2004). Levy (1993) developed an environment-friendly high-capacity dielectric regeneration system capable of autonomously maintaining the quality of the dielectric circulating within the WSEM machine.

1.3 Machining Zone in the SEM Process

Though both WSEM and die-sinking SEM remove material through electrical discharges, there are differences in their operation. Sparking areas in die-sinking SEM are located at the bottom face and the side walls of the tool electrode. This results in an overcut profile on the workpiece, corresponding to the complementary shape of the tool (Fig. 1.6). The overcut profile is known as spark gap and its dimensions are controlled by the SEM process parameters.

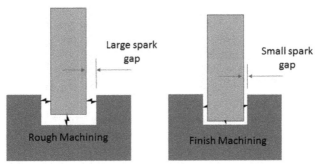

Fig. 1.6: Rough machining and finishing in die-sinking SEM

Sparking in WSEM occurs between the circumference of the semi-circle of the constantly moving wire electrode and the workpiece (Fig. 1.7). A clearance equal to the spark gap is produced on both sides and at the leading face of the electrode. The total width of the electrode and the spark gap is known as 'kerf' and its dimension can be controlled by the WSEM process parameters.

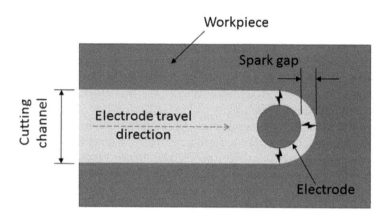

Fig. 1.7: Movement of wire in WSEM

1.4 Important Parameters of SEM Process

Polarity, pulse-on and pulse-off times, pulse frequency, and current are the most important parameters of SEM and have been discussed in the following sections.

1.4.1 Pulse-on Duration

Pulse-on duration or time (T_{on}) in SEM is the time duration for which current flow takes place. It has two components, namely, ignition delay (during which voltage drops from its peak value to servo voltage), and time over which sparking takes place and voltage remains constant at servo voltage (Fig. 1.8). Pulse-on duration governs MRR in SEM – as T_{on} increases, the duration of spark also increases, thereby increasing the energy of the spark; this results in the formation of larger craters on the workpiece surface. But it also increases the surface roughness of the machined workpiece. The value of T_{on} depends on the combination of electrode and workpiece materials. Generally, longer pulse-on durations are used for rough machining and shorter pulse-on durations for finish machining.

1.4.2 Pulse-off Duration

Pulse-off duration or time (T_{off}) (Fig. 1.8) is indicative of the deionization of the dielectric, hence enabling it to regain its insulative characteristics and flushing of the products of machining from the IEG. A smaller T_{off} increases the machining efficiency and the MRR in SEM, but it also yields a poor surface finish due to inefficient flushing; this increases the chances of deposition of the molten material again on the machined surface. A larger T_{off} leads to efficient flushing, thus improving the quality of the machined surface. However, a very small T_{off} compromises the efficiency and the surface quality.

1.4.3 Pulse Frequency

Pulse frequency is the reciprocal of the sum of pulse-on and pulse-off durations, whereas duty cycle is the ratio of pulse-on duration to the sum of pulse-on and pulse-off durations. Generally, a low pulse frequency (Fig. 1.9a) is used for rough machining applications in order to maximise the MRR, while a high pulse frequency (Fig. 1.9b) is preferred for finishing applications so as to reduce the surface roughness of the resulting workpiece.

1.4.4 Current

Current used in SEM and other processes based on SEM characterises the amount of power used. The difference between the peak and the average values of current should be as low as possible (Fig. 1.8) for a workpiece surface of better quality. The magnitude of current in SEM-based processes is governed by the type of application (rough or finish) and the surface area to be machined. Generally, high values of current are used for rough machining and/or machining of large surfaces, and smaller currents are used for finishing and/or machining of smaller surfaces.

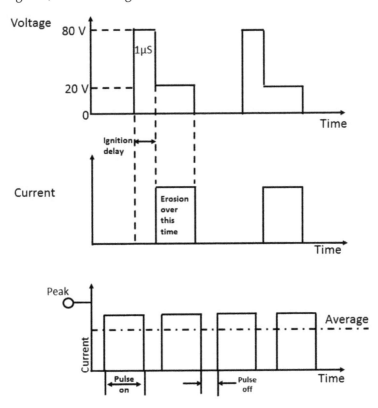

Fig. 1.8: Typical profiles of DC voltage and current used in an SEM process

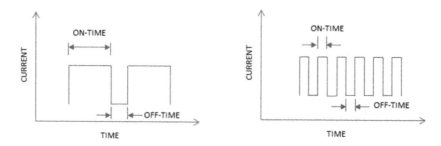

Fig. 1.9: Type of DC pulses used in SEM for (a) roughing applications and (b) finishing applications

1.4.5 Polarity

Since spark erosion in SEM removes material from both the tool and the workpiece, the polarity of the tool and that of the workpiece depend on their relative size and the values of MRR and tool wear rate (TWR). Generally, an SEM process primarily seeks to maximize cathode erosion or anode erosion, depending on the polarity. When TWR is less than MRR, the tool electrode is made into the cathode (negatively charged) and the workpiece acts as the anode (positively charged); these roles are reversed in applications where TWR is more than MRR.

Figure 1.10 shows the variation of erosion rates of cathode and anode as a function of pulse-on duration as developed by DiBitonto et al. (1989). Their research concluded that maximum cathode erosion and minimum anode erosion occur at a pulse-on duration of 30 μs. This is particularly important with reference to die-sinking SEM wherein the tool needs to be protected against significant wear. Hence, in die-sinking SEM, the workpiece is made as the cathode and the tool serves as the anode, and an attempt is made to remove most of the material from the cathode (workpiece) and not from the anode (tool). WSEM uses a very thin wire to machine a workpiece much larger than the wire. Excessive wear of the wire leads to frequent wire breakage, adversely affecting the surface quality and the process economics, thus making it crucial to minimize the wear. Consequently, the wire is used as a cathode and the workpiece as an anode, and a pulse-on duration of 3 μs is required to maximize the erosion (Fig. 1.10).

1.5 Dielectric Fluid

Dielectric fluid is a very important constituent of SEM as it makes spark erosion happen, provided that the required IEG is maintained, while any physical contact between the workpiece and the tool electrode is avoided. It acts as an insulator (does not allow flow of current) until its breakdown

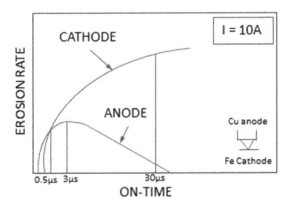

Fig. 1.10: Variation of erosion rates of anode and cathode with pulse-on time
(DiBitonto et al. 1989)

voltage is reached; at breakdown voltage, ionization makes it electrically conducting. The duration of dielectric ionization corresponds to the pulse-on time during which the spark and, consequently, material removal occur. The duration of deionization of the dielectric fluid (dielectric returning to being an electrical insulator) corresponds to the pulse-off time during which the spark-eroded material is flushed out of the IEG. The value of spark frequency can range from 2,000 to 500,000 sparks per second (Elman C. J. 2001). The dielectric fluid aids in maintaining a consistent gap between the tool electrode and the workpiece, while also playing an important role in the removal of the eroded material away from the IEG. Typically, die-sinking SEM uses hydrocarbon oil as a dielectric, whereas WSEM uses deionized water.

1.5.1 Dry SEM

Dry SEM uses a highly pressurized gas or air (instead of hydrocarbon oil) as the dielectric, which is fed in the form of a thin walled pipe through the tool electrode. The gas or air aids the SEM process by continuously removing the debris from the IEG. Its working principle has been illustrated in Fig. 1.11 (Zhang et al. 2002).

Dry SEM was developed to reduce the environmental impacts of using hydrocarbon oils as a dielectric (Abbas et al. 2007). Yu et al. (2004) compared the performance of SEM using a hydrocarbon oil with dry SEM in terms of MRR and electrode wear, and reported the advantages of dry SEM. They found that dry SEM results in a six times higher MRR, while electrode wear is barely 30% of that resulting from SEM using a hydrocarbon oil. Zhanbo et al. (2006) investigated the influence of the rotational speed of the tool electrode, the depth of cut, gas pressure, pulse-on time and pulse-off time on MRR and tool wear in dry SEM. They identified the optimum value of pulse duration as 25 μs, which maximized the MRR and minimized the tool wear.

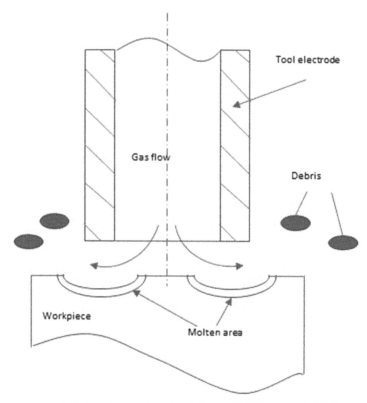

Fig. 1.11: Working principle of dry SEM (Zhang et al. 2002)

Based on their research, Kunieda et al. (2005) made the following conclusions about dry SEM:

- Wear of the tool electrode is negligible for any pulse duration
- Gas supply can be modified to align with the machining application
- The resulting residual stress is small because of a very thin recast layer
- IEG is narrower for dry SEM than for conventional SEM
- Dry SEM is possible in vacuum as long as there is gas flow
- Machines for dry SEM are more compact due to the absence of a storage tank and a supply and recirculation system for the liquid dielectric

1.6 Micro-versions of Spark Erosion Machining

Different micro-versions of SEM and the concept of taking assistance of other sources to enhance the performance of the SEM process have been evolved over time in order to meet the growing demands of additional product functionality. Important examples include micro-SEM (μ-SEM), micro-WSEM, micro-die sinking SEM (μ-DSSEM), micro-spark erosion drilling (μ-SED), micro-spark erosion milling (μ-SEML), and wire spark erosion grinding (WSEG).

Originally, μ-SEM was developed to manufacture holes larger than 200 μm in diameter, in the metallic foils. It uses a tool electrode in the form of a tubular rod with dielectric fluid pumped through it to enhance the flushing within the working gap (Meeusen 2003). After the initial investigations, research interests in this field started fading till late 1980s, when Masuzawa et al. (1985) rediscovered this process by manufacturing very thin tool electrodes on-the-machine by using the WSEG process. These very thin electrodes were meant to be used as micro-pins to manufacture micro-holes. Further development in WSEG took place rather quickly, resulting in cylindrical tool electrodes less than 3 μm in diameter being produced. Tool electrode generation and re-generation is now considered a key enabling technology for stimulating the revival of μ-SEM, for it not only allows process scale down but also minimizes electrode wear. Hence, research interest in and applications of μ-SEM have been increasing continuously. It has thus become an effective and flexible process for the manufacture of complex 3D microstructures, cooling air channels in aerospace turbine blades, tooling inserts for micro-injection molding, micro-filters and micro-fluidic devices, micro-parts for watches, keyhole surgery, housings for micro-engines, and tooling inserts for the fabrication of micro-filters and micro-fluidics devices (Rees et al. 2007). Figure 1.11 illustrates micro-features manufactured by μ-SEM. Furthermore, μ-SEM can be combined with other processes to develop hybrid processes to expand the applications of μ-SEM (Ho et al. 2004).

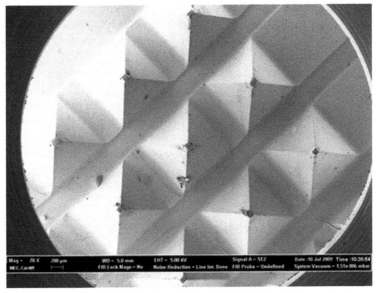

Fig. 1.10 Micro-features machined by μ-SEM

1.6.1 Tool Wear in μ-Spark Erosion Based Processes

Tool wear in μ-SEM is relatively high and cannot be neglected (Yu et al. 1998a, Bigot et al. 2005). Consequently, even the tool wear ratio (the ratio of volumetric wear of the tool electrode to that of the workpiece) is also high. This modifies the μ-SEM process as electrode wear changes the sparking area, thus affecting the accuracy and the quality of the machined part in turn. Several techniques have been applied to minimise the effects of tool electrode wear in μ-SEM and processes derived from μ-SEM.

In micro-SED, the electrode wears in the process of manufacturing blind holes. As a result, while eroding the material to a predetermined depth, the real depth of the hole becomes significantly smaller. One solution to this problem is to repeat the process a number of times with new or regenerated micro-electrodes until the required depth is obtained. This is called the *multiple electrode strategy* (Meeusen 2003). The main drawback of this strategy is that it is time consuming and that it is also difficult to predict the number of electrodes required.

The problems associated with electrode wear are more difficult to address during the manufacture of complex 3D micro-cavities as the wear can be too severe or the electrode geometry may be very difficult to manufacture. The use of micro-SE milling with electrodes of simple shapes has been proposed in the past to solve this problem by using the concept of uniform tool wear (Yu et al. 1998a, 1998b). Another alternative is to use layer-by-layer machining strategy based on the estimation of wear ratio for compensating the electrode wear during the machining of each layer by providing constant electrode feeding in the Z-direction. This requires a very accurate estimation of tool wear as any error therein has a cumulative effect through the layers. Many researchers (Yu et al. 2003, Mohri et al. 1995, Narasimhan 2005) have modelled tool electrode wear, but the accuracy of the proposed models still needs to be verified and improved (Bissacco et al. 2010). Since sparking conditions in μ-SE milling can be application or material dependent and do not remain constant generally, the adoption of machining strategies to compensate the electrode wear is difficult to implement in practice. The main drawback of applying the previously presented wear compensation methods is that they rely highly on the accuracy of the wear estimation models they employ. Therefore, they either underestimate or overestimate the degree of electrode wear, hence resulting in machining inaccuracies. Further developments in compensation methods are expected with improvements in wear estimation models.

1.6.2 Generation of Tool Electrode for μ-SEM

Generation and regeneration of electrodes on-the-machine as envisaged by Masuzawa et al. (1985) remains a key enabling technology for the successful use of μ-SE milling to manufacture complicated microstructures. Tool electrodes for SEM can be generated by *electrode dressing*, which requires intentional wearing of the tool electrode against the objective of minimizing tool wear in the conventional SEM process. Conditions for causing the high wear in μ-SEM include negative polarity, excessive current, high pulse-on times, and high capacitance (Guitrau 1997). Generally, electrodes are dressed through one or a combination of the following processes (Wong et al. 2003, Ho et al. 2004, Ho and Newman 2003, Masuzawa et al. 1985, Mohri et al. 2003):

- Sacrificial block dressing
- Wire spark erosion grinding
- Rotating sacrificial disk

1.6.2.1 Sacrificial Block Dressing

In sacrificial block dressing, also referred to as reverse SEM, a sacrificial conductive block is used to purposely wear the tool electrode, with both of being immersed in a dielectric fluid. The tool electrode is rotated at approximately 500 rpm and the block is brought near the tool electrode. A discharge of sparks causes the dressing of the tool electrode. In this process, the sacrificial conductive block is transformed into the tool electrode and the tool electrode is transformed into the workpiece. Figure 1.13 depicts the principle of this process. Adjustments can be made in the SEM control settings to provide the conditions for the highest wear possible (Guitrau 1997).

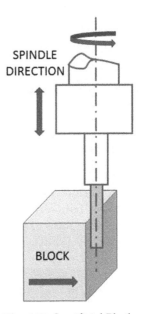

Fig. 1.13: Sacrificial Block dressing (Lim et al. 2003)

Another version of this process was conceived by Yamazaki et al. (2004) by providing a negative polarity to a rod electrode, and rotating and feeding it into a plate, thus making a hole in it. Consequently, after the rod returns to its initial position, its axis is off-centred from the centre of the hole; further, its polarity is reversed, and the rod electrode (with or without rotation) is fed into the plate again. A straight rod electrode can be achieved by this process if the hole does not wear (Yamazaki et al. 2004).

1.6.2.2 Rotating Sacrificial Disk Dressing

In the rotating sacrificial disk dressing process, a continuously rotating cylindrical sacrificial disk of 60 mm diameter and 0.5 mm thickness is used to wear the rotating tool electrode (Fig. 1.14). Both the tool electrode and the sacrificial disk are immersed in a dielectric fluid, and the tool electrode is rotated at approximately 500 rpm while the sacrificial disk is rotated at about 90 rpm. Ultimately, the occurrence of spark discharges between the tool electrode and the rotating disk causes the dressing to occur.

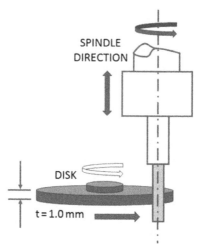

Fig. 1.14: Rotating Sacrificial Disk dressing (Lim et al. 2003)

1.6.2.3 Wire Spark Erosion Grinding (WSEG)

During WSEG, a continuously running wire moving at a speed of approximately 3-5 mm/s is used to wear the electrode (Lim et al. 2003), both immersed in a dielectric fluid. Figure 1.15 illustrates the working principle of WSEG. The tool electrode is rotated at approximately 500 rpm, and spark discharges between the wire and the tool electrode allows the dressing of the tool electrode to occur. Figure 1.16 shows the probe for a micro-CMM machined by WSEG.

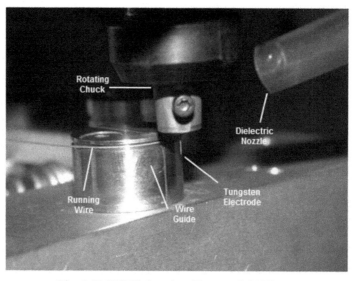

Fig. 1.15: WSEG dressing (Rees et al. 2007)

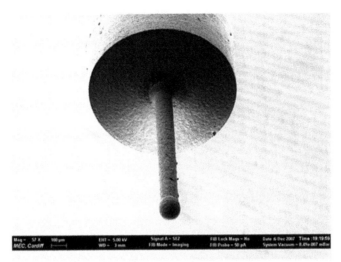

Fig. 1.16: Probe for micro-CMM machined by WSEG process

1.6.3 Power Generators for μ-Spark Erosion Based Processes

Initially, power generators of conventional die-sinking SEM were used in μ-SEM. Two types of pulse generators, namely, relaxation circuit (RC) pulse generators (Fig. 1.17) and transistor type pulse generators (Fig. 1.18) are generally used in conventional SEM (Ho et al. 2004, Han et al. 2004). However, fabrication of parts smaller than several microns requires the lowering of of the pulse energy (<1 μJ) supplied into the gap between the workpiece and the electrode (Masuzawa 2000). This means that finish machining by μ-SEM requires pulse durations of several nanoseconds (Juhr et al. 2004). Since an RC pulse generator can generate such smaller discharge energies simply by decreasing the capacitance in the circuit, it is widely deployed in μ-SEM. Smaller discharge energies generally result in a better quality surface finish with this type of generator (Jahan et al. 2009). However, an RC pulse generator suffers from the following demerits:

- Extremely low MRR due to a low discharge frequency as most of the time is spent on charging of the capacitor
- Difficulty in achieving a uniform surface finish, for the discharge energy varies depending on the electrical charge stored in the capacitor before dielectric breakdown
- Thermal damage to the workpiece if the dielectric strength is not recovered after the previous discharge as the current continues to flow through the same plasma channel in the gap without charging the capacitor

Transistor type pulse generators, widely used in the conventional SEM process, have some advantages: (i) they provide a higher MRR due to their

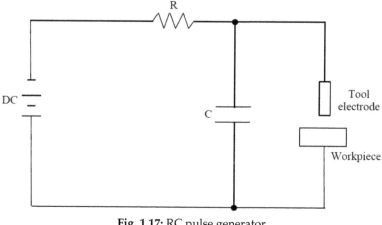

Fig. 1.17: RC pulse generator

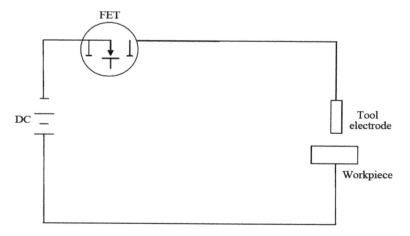

Fig. 1.18: Transistor type pulse generator

high discharge frequency, which obviates the need to charge a capacitor, and (ii) pulse duration and discharge current can be arbitrarily changed, depending on the machining characteristics required. This implies that the use of a transistor type pulse generator in μ-SEM can improve the MRR significantly due to an increase in the discharge frequency.

It may be noted that machines for μ-die-sinking SEM, μ-WSEM and WSEG now fall into the category of applications of μ-SEM.

1.7 Ultrasonic Assistance in SEM

The assistance of ultrasonic vibrations provided to the tool electrode during the SEM process has been used to expand SEM applications and to enhance

SEM machining performance (Abbas et al. 2007). Ultrasonic vibrations have been found to improve the circulation of the dielectric, which facilitates debris removal from the discharge channel. Zhang et al. (1997) proposed a hybrid system that created an ultrasonic frequency using a DC power supply. This produced motion between the tool and the workpiece, hence combining the benefits of ultrasonic machining as well as SEM.

1.8 Surface Integrity Produced by SEM

Characterizing the surface integrity of a surface or a product after its machining is critical in the study of the undesired effects associated with that machining process and for broadening its applications. Particularly, it is important to study the influence of the machining process and the processing conditions on the microstructure and the mechanical properties of the machined surface or product. It is also essential for satisfying the constantly growing requirements of improved functional integration, longevity and reliability of both existing and emerging machining processes and their products.

Field and Kahles (1964) were the first to introduce the concept of surface integrity in a technical sense by defining it as the inherent or enhanced condition of a surface manufactured by any machining process or other surface generation processes. Their subsequent comprehensive review (Fields and Kahles 1971) on surface integrity as particularly encountered in machined components is among the first published literature in this area, and this work emphasized the nature of metallurgical alterations occurring in the surface layers of various materials during material removal processes. Typical surface alterations can be classified as mechanical damages (plastic deformation, micro-cracking, residual stresses, hardness variations), thermal damages (heat-affected zone (HAZ), recrystallization, phase transformation, tears and laps related to built-up edge formation), and chemical damages (inter-granular attack, inhomogeneity, alloy depletion). Subsequently, Field et al. (1972) presented a detailed description of methods for surface integrity inspection, along with an experimental procedure for assessing surface integrity parameters. This involved three different levels of surface integrity data sets to study and evaluate the characteristic features of machined surfaces. Their pioneering contributions gained worldwide recognition and timeless value, further leading to the establishment of an American National Standard for surface integrity (American National Standard on Surface Integrity 1986, Jawahir et al. 2011).

The spark and plasma channels generated during the SEM process create craters on the surface and a thermal wave propagates through the material, resulting in an HAZ on the sub-surface and a recast layer on the surfaces of the components manufactured by SEM (Qu et al. 2002b), thereby modifying their surface integrity (Fig. 1.19). Considering the process-material interactions as a result of the plasma channels generated between the tool electrode and the workpiece (Wong et al. 2003) at the end of the pulse-on time, the molten material is partially ejected and vaporized from the surface by vapor and plasma pressure, but a part of the molten material also remains near the surface due to surface tension forces. The heat quickly dissipates into the bulk of the material and a recast layer is formed due to the re-solidification of the remaining liquid material on the surface. This re-solidification/recast layer is typically very fine grained and found to have a hardness which is twice as hard as the bulk material when machining W300 ferritic steel (Cusanelli et al. 2004). Normally, this layer is subjected to surface tensile stresses, localized hardening, micro-cracking, porosity and grain growth. Also, it may be alloyed with carbon as a side product of dielectric ionization during spark discharge, or with the material transferred from the tool (Ramasawmy et al. 2005). Immediately beneath the recast or white layer lies the HAZ in which heat is not high enough to cause melting, yet sufficiently high to induce micro-structure transformation in the material (Rebelo et al. 1998). Generally, the recast layer that is formed on the surface as a result of SEM is considered as a major drawback of SEM, thus limiting the use of SEM to some critical applications in aerospace, biomedical, automobile and micro-electronics industries (Aspinwall et al. 2008).

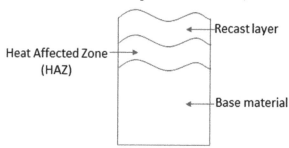

Fig. 1.19: Resulting surface modification by SEM

References

Abbas, N.M., Solomon, D.G. and Bahari, M.F. 2007. A review on current research trends in electrical discharge machining (EDM). International Journal of Machine Tools and Manufacture, 47(7-8): 1214-1228.

Arthur, A., Dickens, P.M. and Cobb, R.C. 1996. Using rapid prototyping to produce electrical discharge machining electrodes. Rapid Prototyping Journal, 2(1): 4-12.

Aspinwall, D.K., Soo, S.L., Berrisford, A.E. and Walder, G. 2008. Workpiece surface roughness and integrity after WSEM of Ti-6Al-4V and Inconel 718 using minimum damage generator technology. CIRP Annals – Manufacturing Technology, 57(1): 187-190.

Bayramoglu, M. and Duffill, A.W. 1994. Systematic investigation on the use of cylindrical tools for the production of 3D complex shapes on CNC SEM machines. International Journal of Machine Tools and Manufacture, 34(3): 327-339.

Bigot, S., Ivanov, A. and Popov, K. 2005. A study of the micro SEM electrode wear. Proceedings of 1st International Conference on Multi-Material Micro Manufacture, pp. 355-358.

Bissacco, G., Valentincic, J., Hansen, H.N. and Wiwe, B.D. 2010. Towards the effective tool wear control in micro-SEM milling. The International Journal of Advanced Manufacturing Technology, 47: 3-9.

Bleys, P.H. and Kruth, J.P. 2001. Machining complex shapes by numerically controlled EDM. International Journal of Electrical Machining, 6: 61-69.

Cusanelli, G., Hessler-Wyser, A., Bobard, F., Demellayer, R., Perez, R. and Flukiger, R. 2004. Microstructure at submicron scale of the white layer produced by SEM technique. Journal of Materials Processing Technology, 149(1-3): 289-295.

DiBitonto, D.D., Eubank, P.T., Patel, M.R. and Barrufet, M.A. 1989. Theoretical models of the electrical discharge machining process – I: A simple cathode erosion model. Journal of Applied Physics, 66(9): 4095-4103.

Elman, C. Jameson 2001. Electrical Discharge Machining, Society of Manufacturing Engineers

Field, M. and Kahles, J.F. 1964. The surface integrity of machined and ground high strength steels. DMIC Report, 210: 54-77.

Field, M. and Kahles, J.F. 1971. Review of surface integrity of machined components. CIRP Annals – Manufacturing Technology, 20(2): 153-163.

Field, M., Kahles, J.F. and Cammett, J.T. 1972. A review of measuring methods for surface integrity. CIRP Annals – Manufacturing Technology, 21(2): 219-238.

Groover, M.P. 2010. Fundamentals of Modern Manufacturing: Materials, Processes and Systems (4th ed.). John Wiley & Sons, New Jersey

Guitrau, E.B. 1997. The SEM Handbook. Chapter 8, page 103. Hanser Gardner Publications, Cincinnati

Han F., Wachi S., Kunieda M. (2004). Improvement of machining characteristics of micro-EDM using transistor type isopulse generator and servo feed system. Precision Engineering, Volume 28, Issue 4, pages 378-385.

Heeren, P.H., Reynaerts, D., Van Brussel, H., Beuret, C., Larsen, O. and Bertholds, A. (1997). Microstructuring of silicon by electro-discharge machining (EDM). Sensors and Actuators A, 61: 379-386.

Ho K.H., Newman S.T. (2003). State of the art electrical discharge machining. International Journal of Machine Tools & Manufacture 43, pages 1287 – 1300.

Ho, K.H., Newman, S.T., Rahimifard, S. and Allen, R.D. 2004. State of the art in wire electrical discharge machining (WEDM). International Journal of Machine Tools and Manufacture, 44(12-13): 1247-1259.

Houman, L. 1983. Total EDM. pp. 5-19. In: Electrical Discharge Machining: Tooling, Methods and Applications. E.C. Jameson (Ed.). Society of Manufacturing Engineers, Dearbern, Michigan.

Huntress, E.A. 1978. Electrical discharge machining. American Machinist, 122(8): 83-98.

Jahan, M.P., Wong, Y.S. and Rahman, M. 2009. A study on the quality micro-hole machining of tungsten carbide by micro-SEM process using transistor and RC-type pulse generator. Journal of Materials Processing Technology, 209(4): 1706-1716.

Jameson, E.C. 2001. Description and Development of Electrical Discharge Machining (EDM). p. 12. *In:* Electrical Discharge Machining. Society of Manufacturing Engineers, Dearbern, Michigan.

Jawahir, I.S., Brinksmeier, E., Saoubi, R.M., Aspinwall, D.K., Outeiro, J.C., Meyerb, D., Umbrellof, D. and Jayala, A.D. 2011. Surface integrity in material removal processes: Recent advances. CIRP Annals – Manufacturing Technology, 60(2): 603-626.

Juhr, H., Schulze, H.P., Wollenberg, G. and Künanz, K. 2004. Improved cemented carbide properties after wire-SEM by pulse shaping. Journal of Materials Processing Technology, 149(1-3): 178-183.

Kalpajian, S. and Schmid, S.R. 2003. Manufacturing Processes for Engineering Materials. Prentice Hall, New Jersey.

Konig, W., Dauw, D.F., Levy, G. and Panten, U. 1988. EDM—future steps towards the machining of ceramics. CIRP Annals – Manufacturing Technology, 37(2): 623-631.

Kruusing, A., Leppavuori, S., Uusimaki, A., Petretis, B. and Makarova, O. 1999. Micromachining of magnetic materials. Sensors and Actuators, 74(1-3): 45-51.

Kunieda, M., Lauwers, B., Rajurkar, K.P. and Schumacher, B.M. 2005. Advancing SEM through fundamental insight into the process. CIRP Annals – Manufacturing Technology, 54(2): 599-622.

Lee, T.C., Zhang, J.H. and Lau, W.S. 1998. Machining of engineering ceramics by ultrasonic vibration assisted SEM method. Materials and Manufacturing Processes, 13(1): 133-146.

Levy, G.N. 1993. Environmentally friendly and high-capacity dielectric regeneration for wire EDM. CIRP Annals – Manufacturing Technology, 42: 227-230.

Levy, G.N. and Wertheim, R. 1988. EDM-machining of sintered carbide compacting dies. CIRP Annals – Manufacturing Technology, 37(1): 175-178.

Lim, H.S., Wong, Y.S., Rahman, M. and Edwin, L.M.K. 2003. A study on the machining of high-aspect ratio micro-structures using micro-EDM. Journal of Materials Processing Technology, 140: 318-325.

Livshits, A.L. 1960. Introduction. *In:* Electro-erosion Machining of Metals. Department of Scientific & Industrial Research, Butterworth & Co., London.

Luo, Y.F., Chen, C.G. and Tong, Z.F. 1992. Investigation of silicon wafering by wire EDM. Journal of Material Science, 27(21): 5805-5810.

Madou, M.J. 2001. Fundamentals of Microfabrication. CRC Press, Boca Raton.

Masuzawa, T. 2000. State of the art of micro machining. CIRP Annals – Manufacturing Technology, 49(2): 473-488.

Masuzawa, T., Fujino, M. and Kobayashi, K. 1985. Wire electro-discharge grinding for micro-machining. CIRP Annals – Manufacturing Technology, 34(1): 431-434.

Meeusen, W. 2003. Micro-electro-discharge: Technology, computer-aided design & manufacturing and applications. PhD thesis, Department of Mechanical Engineering, Leuven.

Mohri, N., Fukuzawa, Y., Tani, T., Saito, N. and Furutani, K. 1996. Assisting electrode method for machining insulating ceramics. CIRP Annals – Manufacturing Technology, 45(1): 201-204.

Mohri, N., Suzuki, M., Furuya, M. and Saito, N. 1995. Electrodes wear process in electrical discharge machining. CIRP Annals – Manufacturing Technology, 44(1): 165-168.

Mohri, N., Takezawa, H., Furutani K., Ito, Y. and Sata, T. 2003. A new process of additive and removal machining by SEM with a thin electrode. CIRP Annals – Manufacturing Technology, 49(1): 123-126.

Muller, F. and Monaghan, J. 2000. Non-conventional machining of particle reinforced metal matrix composite. International Journal of Machine Tools Manufacturing, 40(9): 1351-1366.

Narasimhan, J., Yu, Z. and Rajurkar, K.P. 2005. Tool wear compensation and path generation in micro and macro EDM. Journal of Manufacturing Processes, 7(1): 75-82.

Qu, J., Shih, A.J. and Scattergood, R.O. 2002b. Development of the cylindrical wire electrical discharge machining process, Part 2: Surface integrity and roundness. Journal of Manufacturing Science and Engineering, Transactions of the ASME, 124(3): 708-714.

Rajurkar, K.P., Levy, G., Malshe, A., Sundaram, M.M., McGeough, J., Hu, X., Resnick, R. and DeSilva, A. 2006. Micro and Nano Machining by Electro-Physical and Chemical Processes. CIRP Annals – Manufacturing Technology, 55(2): 643-666.

Ramasawmy, H., Blunt, L. and Rajurkar, K.P. 2005. Investigation of the relationship between the white layer thickness and 3D surface texture parameters in the die sinking SEM process. Precision Engineering, 29(4): 479-490.

Rebelo, J.C., Dias, A.M., Kremer, D. and Lebrun, J.L. 1998. Influence of SEM pulse energy on the surface integrity of martensitic steels. Journal of Materials Processing Technology, 84(1-3): 90-96.

Rees, A., Dimov, S., Ivanov, A., Herrero, A. and Uriarte, L. 2007. Micro-electrode discharge machining: Factors affecting the quality of electrodes produced on the machine through the process of wire electro-discharge machining. Proceedings of the Institution of Mechanical Engineers, Part B: Journal of Engineering Manufacture, 221: 409-418.

Rhoney, B.K., Shih, A.J., Scattergood, R.O., Akemon, J.L., Grant, D.J. and Grant, M.B. 2002. Wire electrical discharge machining of metal bond diamond wheels for ceramic grinding. International Journal of Machine Tools and Manufacture, 42(12): 1355-1362.

Rzmasawmy, H. and Blunt, L. 2002. 3D surface topography assessment of the effect of different electrolytes during electrochemical polishing of SEM surfaces. International Journal of Machine Tools and Manufacture, 42: 567-574.

Sanchez, J.A., Cabanes, I., Lopez de Lacalle, L.N. and Iamikiz, A. 2001. Development of optimum electrodischarge machining technology for advanced ceramics. International Journal of Advanced Manufacturing Technology, 18(12): 897-905.

Storr, M. 2019, Dielectrics for electric discharge machining, OEL-HELD GmbH Publication.

Stovicek, D.R. 1993. The state-of-the-art EDM Science. Tooling and Production, 59(2): 42.

Toren, M., Zvirin, Y. and Winograd, Y. 1975. Melting and Evaporation Phenomenon during Electrical Erosion. Journal of Heat Transfer, 576-581.

Valentincic, J., Brissaud, D.B. and Junkar, M. 2006. EDM process adaptation system in toolmaking industry. Journal of Materials Processing Technology, 172: 291-298.

Webzell, S. 2001. That first step into EDM. Machinery, 159: 41.

Wong, Y.S., Rahman, M., Lom, H.S., Nan, H. and Ravi, N. 2003. Investigation of micro-SEM material removal characteristics using single RC-pulse discharges. Journal of Materials Processing Technology, 140(1-3): 303-307.

Yamazaki, M., Suzuki, T., Mori, N. and Kunieda, M. 2004. SEM of micro-rods by self-drilled holes. Journal of Materials Processing Technology, 149: 134-138.

Yan, B.H., Wang, C.C., Liu, W.D. and Huang, F.Y. 2000. Machining characteristics of Al_2O_3/6061Al composite using rotary SEM with a disklike electrode. The International Journal of Advanced Manufacturing Technology, 16(5): 322-333.

Yeo, S.H., Kurnia, W. and Tan, P.C. 2008. Critical assessment and numerical comparison of electro-thermal models in EDM. Journal of Materials Processing Technology, 203(1-3): 241-251.

Yu, Z.B., Jun, T. and Masanori, K. 2004. Dry electrical discharge machining of cemented carbide. Journal of Materials Processing Technology, 149: 353-357.

Yu, Z.Y., Kozak, J. and Rajurkar, K.P. 2003. Modelling and Simulation of Micro EDM Process. CIRP Annals – Manufacturing Technology, 52(1): 143-146.

Yu, Z.Y., Masuzawa, T. and Fujino, M. 1998a. Micro-SEM for three-dimensional cavities – development of uniform wear method. CIRP Annals – Manufacturing Technology, 47(1): 169-172.

Yu, Z.Y., Masuzawa, T. and Fujino, M. 1998b. 3D Micro-SEM with simple shape electrode. International Journal of Electrical Machining, 3: 7-12.

Zhanbo, Y., Takahashi, J., Nakajima, N., Sano, S., Karato, K. and Kunieda, M. 2006. Feasibility of 3-D surface machining by dry EDM. http://www.sodic.co.jptechimgarticle_s02.pdfS (Downloaded on May 1 2006).

Zhang, J.H., Lee, T.C., Lau, W.S. and Ai, X. 1997. Spark erosion with ultrasonic frequency. Journal of Materials Processing Technology, 68: 83-88.

Zhang, Q.H., Zhang, J.H., Deng, J.X., Qin, Y. and Niu, Z.W. 2002. Ultrasonic vibration electrical discharge machining in gas. Journal of Materials Processing Technology, 129: 135-138.

PART II

Machining of Biomedical Components, Shape Memory Materials, and Aerospace Alloys

Wire Spark Erosion Machining of Shape Memory Materials

Himanshu Bisaria and Pragya Shandilya*
Department of Mechanical Engineering, MNNIT Allahabad, UP, India

2.1 Introduction to Shape Memory Materials

Metals and alloys have been used as structural materials for centuries. With the advancement in the field of material science, and the increasing restrictions in logistics and space, engineers and researchers have been continuously developing high-performance materials for various applications (Ghosh et al. 2013). In many cases, the unending goal of engineers is to enhance the efficiency of products and to reduce their weight. To achieve these goals, they have been looking for replacements of multi-component and multi-material systems with high-performance multi-functional and light-weight materials. One subgroup of multi-functional materials is active materials with *sensing* (converting mechanical input to non-mechanical output) and *actuation* (converting non-mechanical input to mechanical output) capabilities (Kumar and Lagoudas 2008). Active materials are further classified into two categories: (i) materials with direct coupling (for example, piezoelectric ceramics and polymers, piezomagnetic or magnetostrictive ceramics, shape memory materials (SMM), and magnetic-SMM), and (ii) materials with indirect coupling (for instance, magneto-rheological fluids (MRF) and electro-rheological fluids (ERF)) (Momoda, 2004). Shape memory materials (SMM) are active multi-functional smart materials with shape recovery ability (SRA), along with phase transformations (solid-to-solid) induced by external non-mechanical stimuli such as changes in temperature, stress or magnetic field (Huang et al. 2010), and respond to them reversibly. The history of SMM dates back to 1932, when Arne Olander discovered them. However, the term 'shape memory' was first described by Vernon a few years later in 1941. Dr. William Buehler's group at the US Naval Ordinance Laboratory, in 1963, recorded the presence of an exceptional *shape memory*

*Corresponding author: pragya20@mnnit.ac.in

phenomenon in the 50-50 alloy of nickel-titanium (Ni-Ti); since then, Ni-Ti alloys came to be referred to as NITINOL (Jani et al. 2014). Under specific conditions, SMM have the ability to absorb and disperse mechanical energy during reversible hysteretic changes in their shape when exposed to cyclic mechanical loading. These exceptional features of SMM make them suitable for applications involving absorption of impact and vibration, sensing and actuation, and to recover seemingly permanent strains. SMM have unique properties, namely, *shape memory effect, biocompatibility, superelasticity (SE)* or *pseudoelasticity*, and higher resistance to corrosion and wear. Hence, SMM are used in a variety of applications, like satellites, robotics, space engineering, aerospace engineering, biomedical engineering, astronomy, micro- and nano-electro mechanical systems (MEMS and NEMS) (Hartl and Lagoudas 2007).

2.1.1 Phase Transformations in SMM

SMM exist in austenite and martensite phases with three distinct crystal structures: austenite in face centered cubic (FCC), twinned martensite in alternatively shear platelet structure, and detwinned or deformed martensite. Transformations among these crystal structures take place in response to changes in the external environment, such as stress, temperature, pressure, or magnetic field (Kamila 2013). The high-temperature phase with FCC or body centered cubic (BCC) crystal structure, referred to as the parent phase or the austenitic phase, is cooled below the martensitic transformation temperature, for transforming into the twinned martensite phase. This phase has lower crystallographic symmetry. SMM are easier to deform in this phase, and such deformations lead to the deformed martensite phase. Figure 2.1a shows the microstructural transformation of SMM during thermoelastic phase transformation between austenite and martensite phases (Sun and Huang 2009). In the figure, A_s, A_f, M_s and M_f are the temperatures associated with the start and the end points of formation of austenite and martensite phases respectively. When subjected to heat, SMM transform from martensitic phase (M) to austenitic phase (A). On further heating beyond the A_s temperature, contraction takes place and SMM are transformed into austenite phase (original phase) (Mihálcz 2001). In crystalline materials, two types of phase transformations are possible: diffusional and diffusionless (or displacive). The martensitic transformation belongs to the category of displacive or diffusionless transformation and is characterized by well-coordinated shear dominant atomic displacement with volume change during the phase transformation being very small.

Calorimetric or electrical resistivity measurements can be used to detect phase transformations in SMM. The phenomenon of martensitic phase transformations (A to M or M to A) is related to the absorption and release of latent heat (Shimizu 2011). Heat and temperatures related to phase transformations are commonly determined by the differential scanning calorimeter (DSC) test. A schematic diagram of the DSC graph for SMM showing various transformation temperatures is illustrated in Fig. 2.1b (Kumar and Lagoudas 2008). SMM have two interesting properties, namely, shape memory effect (SME) and superelasticity associated with reversible solid-state diffusionless thermoelastic phase transformation. SME can be defined as the capability of an SMM to memorize its preset shape on the application of heat, whereas superelasticity (SE) or pseudo-elasticity is defined as its ability to recover approximately 8% of the strain and is related to the stress-strain (σ-ε) hysteresis due to mechanical loading and mechanical unloading in isothermal states (Kumar and Lagoudas 2008). The SRA of an SMM can be measured by the SME bending test. The schematic diagram of SME bending test has been presented in Fig. 2.2, wherein the surface bending

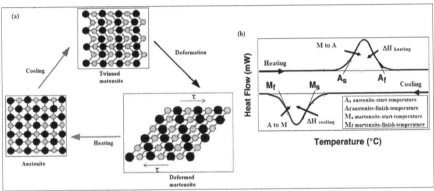

Fig. 2.1: Schematic diagram: (a) microstructural transformation of shape memory material (b) differential scanning calorimeter curve for a typical NiTi-based shape memory material (Sun and Huang 2009 © Springer Nature. Modified reprint with permission)

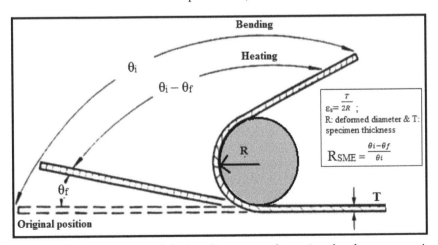

Fig. 2.2: Schematic diagram of the bending test to determine the shape recovering ability of shape memory materials (Lin and Wu 1992 © Elsevier. Reprinted with permission)

strain (ε_s) and the SRA in terms of R_{SME} are shown by means of specimen geometry (Rao et al. 2015).

2.1.2 Types of SMM

Figure 2.3 presents a detailed classification of SMM. Besides metals and alloys, different polymers, ceramics and other materials also exhibit the shape memory effect. Metals and alloys which exhibit SME can be classified into three groups: shape memory alloys (SMA), high-temperature SMA (HTSMA), and magnetic SMA (MSMA) (Kohl 2004, Ma et al. 2010). SMA can be further classified on the basis of the parent material, that is, NiTi-X, Cu-X, Fe-X, Ag-X, Au-X, and Co-X SMM (X representing the other element). Table 2.1 summarizes the properties of NiTi, Cu and Fe-based SMM.

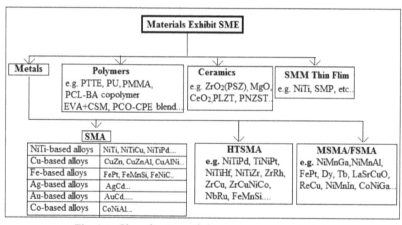

Fig. 2.3: Classification of shape memory materials

HTSMA is an exclusive class of SMA with high phase transformation temperatures (> 100°C) and the capability of actuating under high-temperature conditions (Ma et al. 2010). Addition of ternary elements (Pd, Pt, Hf, Au, and Zr) to NiTi alloys results in HTSMA with phase transformation temperatures in the range of 100 to 800°C. Based on the martensitic transformation ranges, these HTSMA can be further categorized into three groups: (i) HTSMA with martensitic transformation temperatures ranging from 100 to 400°C (e.g. TiNiPd, TiNiPt, NiTiZr, NiTiHf, CuAlNi, CuAlNb, CoAl, NiAl,), (ii) HTSMA with martensitic transformation temperatures ranging from 400 to 700°C (e.g. TiPd, TiAu), and (iii) HTSMA with martensitic transformation temperatures above 700°C (e.g. TiPtIr, TaRu, NbRu). The main restriction with HTSMA is the low value of critical strain for slip and the phase transformation being approximately 3% (Firstov et al. 2006).

MSMA, also called ferromagnetic SMA (FSMA), have the ability of actuation at high frequencies (up to 1 kHz) as the actuation energy is spread through magnetic fields and the relatively slow mechanism of heat transfer

Table 2.1: Properties of different shape memory alloys (Tadalii 1999, Maruyama and Kubo 2011)

Property	NiTi-X based SMA	Cu-X based SMA	Fe-X based SMA
Specific heat (J/kg °C)	450-620	390-440	550
Thermal conductivity (at 20 °C in W/mK)	8.6-18	30-120	8.5
Density (g/cm³)	6.4-6.5	7.1-8.0	7.2-7.5
Latent heat (kJ/kg)	19-32	-	-
Maximum recovery stress (MPa)	500-900	400-700	400
Transformation temperature (°C)	–200 to 200	–200 to 200	–200 to 150
Poisson ratio	-	0.34	0.36
Elongation (%)	-	8-15	16-30
Normal working stress (MPa)	100-130	150-400	200-400
Young's Modulus (GPa) (M and A)	28-83	65-75	140-200

does not restrict it. Comparatively low values of *blocking stress* (at this stress, the magnetic reorientation strains are totally repressed) are a major challenge associated with MSMA. Some examples of FSMA and meta-magnetic SMA (MMSMA) are FeNiCoTi, FeNiCoAl and FeNiMnAl alloys, which have capability to perform at relatively high frequencies (Fujita 2000, Wutting et al. 2001).

2.1.3 Manufacturing of SMM

The different processes used to manufacture SMM can be classified into two categories, namely, casting and powder metallurgy. Figure 2.4 depicts the different manufacturing processes under these two categories. Casting processes are most suitable for large scale manufacturing of NiTi alloys; here the first ingots of Ni and Ti are melted, hot worked and final machined (Jani et al. 2014). Vacuum induction melting (VIM) and vacuum arc melting (VAM) are the most commonly used casting processes for the manufacture of NiTi SMA (Morgan and Broadley 2004). The VIM process has an advantage of yielding a homogeneous chemical composition throughout the ingot. In this process, the constituent metals are melted in vacuum or in an inert atmosphere. It is generally preferred for materials with strong oxidation affinities. The VAM process is used to manufacture ingots of large size. In this process, the problem of contamination from the crucible is eliminated, unlike in the VIM process. Furthermore, either a consumable or a non-consumable electrode is used instead of crucible in this method (Morgan and Broadley 2004). On the other hand, powder metallurgy processes are useful for the production of NiTi components in small volumes, mainly used in biomedical applications for producing porous NiTi components (Miyazaki and Sachdeva 2009).

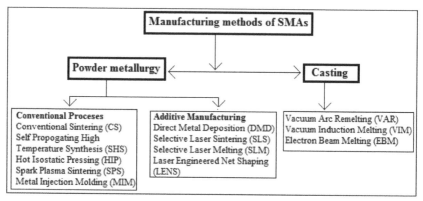

Fig. 2.4: Different processes for manufacturing shape memory materials

2.1.4 Applications of SMM

The major fields of applications of SMM include biomedical, robotics, automotive, aerospace and other important industries (Fig. 2.5) (Sutapun et al. 1998, Hartl and Lagoudas 2007, Song et al. 2006). The special properties of SMM, particularly, kink resistance, excellent MRI compatibility, and corrosion resistance, make them a good choice for various biomedical applications

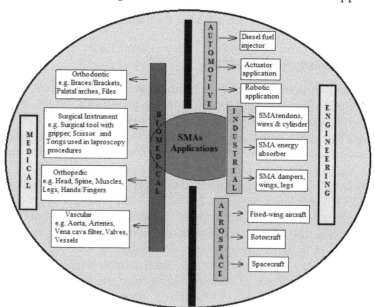

Fig. 2.5: Main applications of shape memory materials

(Morgan 2004, Thompson 2000). SMM are also used in biomedical devices and equipment in cardiology, neurology, orthopedic implants, eye glass frames, micro-electro-mechanical system (MEMS) devices, and actuators (Gil and Planell 1998, Quan and Hai 2015). Actuation applications of SMM require SMM to (i) have the ability to freely revert to their original shape, or (ii) be fully constrained in such a way that they exert large forces on the constraining structure after shape recovery, or (iii) be partially constrained by the surrounding deformable structure. Some practical examples of actuation devices include vascular stents, thermostats, hydraulic pipe couplings, and Coffee maker. Components such as eye-glass frames and arch wires in orthodontics use the superelasticity of SMM. The twinned martensite phase has excellent energy absorbing and fatigue resistance capabilities, making it suitable for applications in vibration dampener and flexible surgical tools for open-heart surgery.

2.2 WSEM of SMM

The same unique properties of SMM, which lend usefulness to them, pose challenges in the creation of engineering products from them. They also cause the use of conventional machining processes (turning, drilling, milling, etc.) to be extremely difficult or even impossible in some cases. Consequently, researchers have started exploring the use of advanced processes for efficient machining of SMM. Although every advanced machining process has its own limitations, wire spark erosion machining (WSEM) has an excellent potential to machine SMM due to its capability of producing near net-shaped and burr-free products with superior surface quality and higher dimensional stability (Gupta and Jain 2014). WSEM can be used to manufacture any intricate 2D and 3D shape in any electrically conducting material through a series of sparks between a generic tool in the form of a thin wire (of 100 to 500 μm diameter) and the workpiece. In the case of micro-WSEM, the diameter of the wire may be less than 20 μm (Bisaria and Shandilya 2015). Slow and continuous feeding of the wire results in the desired cutting or machining path. The relative motion between the workpiece and the wire is computer numerically controlled (CNC). Moreover, an inter-electrode gap of 25 to 50 μm between the wire and the workpiece is constantly maintained by the CNC system (El-hofy 2005). The machine for WSEM consists of four basic components: (i) DC power supply system, (ii) wire feed system, (iii) work positioning system, and (iv) dielectric supply system (Benedict 1987). WSEM uses very high pulse frequencies (of the order of 1 MHz) and pulse-off times, and smaller pulse-on times and current. The function of the wire feeding system is to continuously supply fresh wire under constant tension to the machining area. A constant wire tension is needed to avoid problems such as wire lag, wire vibrations, wire breaks and streaks marks. Arrangements of various pulleys are used to ensure a proper positioning of the wire in order to provide the wire with axial tension and the programmed feed rate. The upper and lower guides are used to drive the wire according to the programmed path. These guides are made of wear resistant material such as sapphire or diamond. The diameter of the guides depends on the diameter of the wire. The WSEM work positioning system usually consists of a 2-axis CNC table the operation of which keeps a constant spark gap between the workpiece and the wire (Jain 2007). It contains two guides of stainless steel with sufficiently threaded holes, and the surface is in a definite position relative in the X-Y plane and placed in the path of wire movement. Figure 2.6 depicts a schematic of the WSEM process.

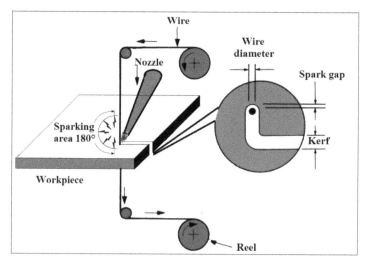

Fig. 2.6: Schematic representation of the WSEM process (Benedict 1987)

Selection of wire material is yet another important aspect in WSEM. The chemical and physical characteristics of the wire affect the performacne of the WSEM process. Generally, a brass wire is used for WSEM. Depending on the production process, a brass wire with suitable tensile strength (490-900 N/mm^2) can be selected. Furthermore, different material coatings can be used on the wire to enhance its different properties. Elongation (%) and tensile strength of different coated wires has been shown in Fig. 2.7. The basic function of the dielectric supply system is to provide a continuous supply of dielectric fliud to the machining zone. Deionized water is the commonly used dielectric in WSEM, due to its unique characteristics of low viscosity, low discharge voltage, high cooling rate, high MRR and no fire hazard (Jameson 2001). The process parameters of WSEM can be categorized into two major groups: electric and non-electric process parameters. The electrical parameters include spark gap voltage or servo-voltage (V_s), pulse peak voltage (V_d), pulse-off duration (T_{off}), pulse-on duration (T_{on}) and peak current (I_p); whereas the non-electric parameters can be further categorized

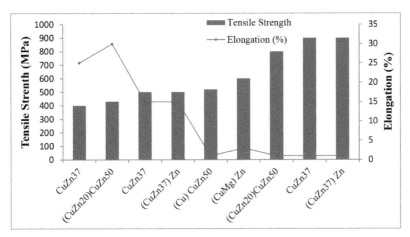

Fig. 2.7: Tensile strength and elongation (%) of different coated wires used in WSEM

into sub-categories on the basis of the dielectric (conductivity, pressure), the wire electrode (material, size, tension, and feed rate) and the workpiece (type of material and dimensions) (Sanchez and Ortega 2009). The value of maximum voltage occurring during the ignition delay period of pulse-on time is referred to as pulse peak voltage 'V_d', and the voltage during spark erosion is referred to as spark gap voltage or servo-voltage 'V_s' (refer to Fig. 1.8).

The WSEM process is generally used for making different types of dies, fabrication of press tools, through-hole machining and precision tooling from electrically conducting but difficult-to-machine materials (like superalloys, titanium alloys, metal matrix composites, and SMM) for applications in aerospace, automotive, robotics and biomedical industries (Benedict 1987, Jain 2007, Sanchez et al. 2009). Garg et al. (2014) studied the effects of V_s, T_{off}, T_{on}, I_p, wire feed rate (WF) and wire tension (WT) on surface roughness (SR) and cutting speed during WSEM of Titanium 6-2-4-2 alloy (which is a near alpha alloy of titanium, aluminum, tin, zirconium and molybdenum) and observed that V_s, I_p, T_{off} and T_{on} were the significant parameters, whereas WF and WT were comparatively less significant. Kumar et al. (2017) observed noticeable influence of I_p and T_{off} on surface roughness of nickel-based super alloys and an increase in surface roughness with increase in T_{on} and I_p, which in turn increased the discharge energy, resulting in overheating and evaporation of the molten workpiece material; this created craters of bigger sizes on the machined surface, thus increasing the surface roughness. During studies of WSEM of SiCp/6061 aluminum metal matrix composites (MMC), it was concluded that the combination of lower values of V_s, T_{off} and wire feed rate, and higher values of T_{on} results in a higher average cutting rate and MRR (Shandilya et al. 2012).

WSEM can be successfully used for efficient machining of SMM to overcome the difficulties associated with conventional machining of SMM, such as the formation of a wide hardened layer on the cutting edge, strain hardening, cyclic hardening, high machining time, and high tool wear (Lin et al. 2000, Weinert et al. 2004). Due to its unique features like better surface finish, ability to machine complicated profiles, and low residual stresses, the WSEM process is generally considered for machining SMM. Bisaria and Shandilya (2019) observed an increase in the MRR of Ni-based SMM with the increase in I_p and T_{on}; it was attributed to an increase in the discharge energy, which resulted in more melting and evaporation of the workpiece material. The machining characteristics and surface integrity of SMM in WSEM have been studied under this section.

2.2.1 Characteristics of WSEMed SMM

MRR, cutting rate, surface roughness and surface integrity are the important measures of performance of WSEM; they are selected during the evaluation of machinability of SMM by WSEM. During studies of machinability of TiNiX ternary SMM (where X = Zr/Cr) by WSEM, it has been observed that the product of melting temperature T and thermal conductivity K of an SMM strongly affects the machining characteristics. SMM with high melting temperatures result in less melting and evaporation, whereas those with higher thermal conductivity lead to higher transfer of discharge energy, thus resulting in a lower cutting rate. The maximum wire feed rate without wire breakage in WSEM has a reverse relationship with the KT product. Therefore, the product KT for an SMM can be treated as a characteristic of WSEM of SMM (Hsieh et al. 2009). Figure 2.8 represents the value of KT for different SMM. Furthermore, like other conductive alloys, the machining characteristics of SMM also depend on the parameters of WSEM.

Higher values of MRR and SR are primarily attributed to higher T_{on} and I_p, and lower T_{off} and V_s; this is due to an increase in the discharge energy and

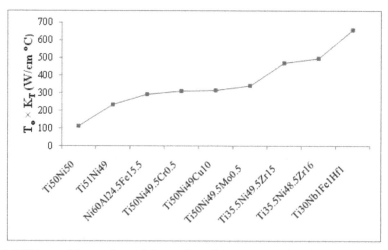

Fig. 2.8: Values of the product of thermal conductivity K and melting temperature T for different shape memory materials

the intensity of spark. Soni et al. (2017) observed that increasing T_{on} increased both MRR and SR of $Ti_{50}Ni_{40}Co_{10}$ SMM, whereas an increase in T_{off} and V_s decreased them. At a fixed value of T_{off} and V_s, Manjaiah et al. (2015a) found the MRR of $Ti_{50}Ni_{40}Cu_{10}$ and $Ti_{50}Ni_{30}Cu_{20}$ SMM during the WSEM process to increase from 4 mm^3/min to 8 mm^3/min, along with an increase in T_{on}. It was observed that at a fixed value of I_p, table feed rate and T_{off}, the MRR of $Ti_{50}Ni_{45}Cu_5$ SMM increased with T_{on}. Furthermore, higher T_{on} also affected the wire feed rate and the impact of the dielectric fluid (Manjaiah et al. 2015b). Surface roughness depends on crater size and discharge energy. An increase in discharge energy not only increases the surface roughness, but also the formation of craters, pockmarks, voids and globules. Moreover, the WSEM process is also affected by wire speed.

2.2.2 Surface Integrity of WSEMed SMM

Surface integrity is the sub-surface condition (up to 0.5 mm below the manufactured surface) of any component manufactured by any process. It includes not only the topological (geometrical) aspects of the manufactured surface but also its physical, chemical, metallurgical, mechanical and biological aspects (Grzesik et al. 2010). The surface integrity aspects (such as surface topography, metallurgical changes, micro-hardness, recast layer and SRA) of SMM having gone WSEM are discussed in this section.

2.2.2.1 Surface Topography

Surface topography invariably depends on the variable process parameters of WSEM. Since material removal in the WSEM process takes place at extremely high temperatures (even reaching up to 10,000°C for a fraction of a second), the surface of $Ni_{50.89}Ti_{49.11}$ SMM after machining consists of many cracks, craters, voids, micro-globules of the melted droplets, and debris (Fig. 2.9). When the effects of T_{off}, T_{on} and V_s on the surface topography of $Ni_{50.89}Ti_{49.11}$ SMM were studied, it was found that at higher discharge energy parameters (high T_{on} and V_s, and low T_{off}), more micro-cracks and micro-globules, and larger craters were formed on the WSEMed surface. Manjaiah et al. (2016) observed blow holes on the WSEMed surface of $Ti_{50}Ni_{40}Cu_{10}$ SMM due to overheated melted material entrapping to the escapement of gas bubbles. Manjaiah et al. (2015a) observed that at lower T_{off}, numerous large craters were present on the WSEMed surfaces of $Ti_{50}Ni_{40}Cu_{10}$ and $Ti_{50}Ni_{30}Cu_{20}$ SMM, whereas fewer wider craters were observed at higher T_{off}. As the crater size depends on the discharge energy, more discharge strikes the workpiece surface with such an increase; this leads to the formation of deeper and larger craters, thus causing higher surface roughness (Manjaiah et al. 2015c).

2.2.2.2 Recast Layer Formation

The subsurface generated after WSEM consists of chemically and metallurgically affected zones. There are three types of layers in these zones:

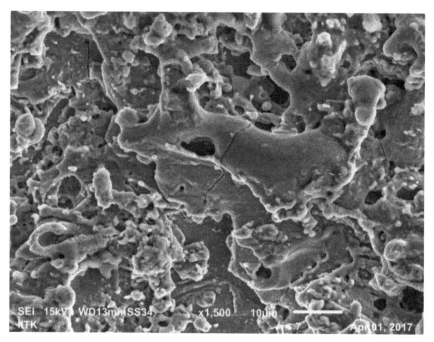

Fig. 2.9: Scanning electron micrograph of $Ni_{50.89}Ti_{49.11}$ SMM machined by the WSEM process (Bisaria and Shandilya 2018 © Taylor and Francis. Reprinted with permission)

recast layer (RL) or white layer, heat affected zone (HAZ), and converted layer. The outermost layer having the recast structure is called recast layer or white layer (white color under a microscope). This recast layer is very hard and brittle in nature, with a hardness value greater than 65 HRC. It is a layer formed due to phase transformations on the surface after machining, the thickness of which depends on the level of discharge energy. In WSEM, owing to the rapid quenching effect, a thin RL is formed near the outer machining zone through the re-solidification of the melted material; this causes transformation of the phases and various microstructures, compared to the bulk material (Grzesik et al. 2010). Beneath the RL lies the HAZ, where heat is not sufficient enough to cause melting or recasting, but ample enough to encourage microstructural transformation. The converted layer lies beneath the HAZ.

During the WSEM of $Ti_{50}Ni_{40}Co_{10}$ SMM, Sonia et al. (2017) observed the formation of a thick RL on the WSEMed surface at lower V_s and higher T_{on}; this was due to the transfer of more discharge energy to the workpiece, causing more melted material to re-solidify on the WSEMed surface. Liu et al. (2016) observed a non-uniform and porous bi-layered structured white layer on the WSEMed surface of nitinol (SE508) SMM, using scanning electron microscopy (SEM). The upper fragment of this white layer contained a solid solution of Cu/Zn/Ni phases, whereas the lower portion consisted of Ti_2O_3 and nitinol austenite phase. Bisaria and Shandilya (2019) noted a decrease in the thickness of RL, with increasing pulse-on duration during WSEM of TiNiX ternary SMM. At higher pulse-on durations, the dielectric fluid has enough impact to wipe out the melted materials and the particles deposited on the WSEMed surface, thus reducing the RL thickness (Hsieh et al. 2009). In another study, Liu et al. (2014) observed a non-uniform and discontinuous porous white recast layer (RL) of thickness 2-8 μm on the machined surface in WSEM of nitinol (SE508) SMM, in the main cut mode. The thickness of this RL considerably reduced during the trim cutting mode (using comparatively lower values of discharge energy). In the first trim cut, the thickness of RL reduced to 1-4 μm, and finally to 0-2 μm in the subsequent trim cut.

Besides metallurgical changes, the chemical changes in a WSEMed surface are also observed due to the reaction of the workpiece with the wire electrode and the dielectric at elevated temperatures. During WSEM, the foreign particles from the wire and the dielectric are transferred to the workpiece. Figure 2.10 represents the EDS analysis of the RL formed on the WSEMed surface of $Ni_{50.89}Ti_{49.11}$ SMM at T_{on} of 105 and 125 μs, others parameters kept constant (T_{off} = 54 μs; V_s = 50 V; WT = 5 N; WF = 8 m/min). It illustrates the transfer of foreign atoms (Cu, Zn, C and O) from the wire electrode and the dielectric to the machined surface in WSEM (Bisaria and Shandilya 2018).

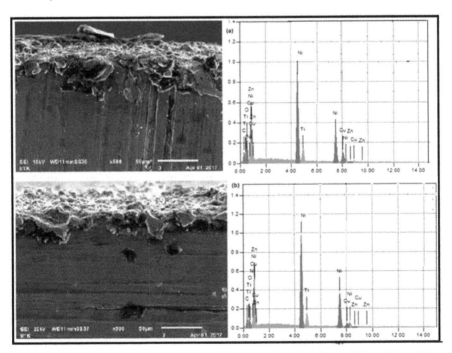

Fig. 2.10: EDS analysis of the recast layer formed on the surface of $Ni_{50.89}Ti_{49.11}$ SMM (T_{off} = 54 μs, V_s = 50 V, WF = 8 m/min, WT = 5 N) (a) T_{on} = 105 μs and (b) T_{on} = 125 μs (Bisaria and Shandilya 2018 © Taylor and Francis. Reprinted with permission)

Liu et al. (2014) studied the surface integrity aspects of rough cutting and finish cutting modes in WSEM of SE508 (50.8% Ni and 49.2%Ti) SMM. They found that the diffusion of foreign elements (Cu, Zn, C and O) in the crater is related to not only the discharge energy but also the pulse frequency; even pulse-on duration may affect the diffusion of these foreign particles. Presence of foreign elements can be validated through an EDS analysis of the recast layer. The RL formed on the WSEMed surface of the $Ti_{50}Ni_{40}Cu_{10}$ SMM consists of Cu, Zn, C and O, besides the base material (Ni and Ti). The atoms of Cu, Zn, C and O from the wire electrode and the dielectric are transferred to the surface at very high temperatures during WSEM (Manjaiah et al. 2015c). However, this recast layer formed near the WSEMed surface does not affect the SRA of SMM at normal bending strains; however, SRA is slightly degraded at higher bending strains. In case of thin plates, this RL formed on the WSEMed surface can be removed by mechanical means or by electrochemical polishing, thus improving the SRA of SMM (Lin et al. 2005, Hsieh et al. 2009).

2.2.2.3 Micro-hardness

Variation of micro-hardness with pulse-on duration for the WSEMed surface of $Ni_{50.89}Ti_{49.11}$ SMM has been presented in Fig. 2.11. It can be observed that surface hardness near the outer machining zone is several times higher than the hardness of the base material ($Ni_{50.89}Ti_{49.11}$ SMM). Due to the 'quenching effect' by sudden heating and cooling of the surface, and the formation of oxides, carbides and over-tempered martensite, the hardness of the WSEMed surface increases (Bisaria and Shandilya 2019). Manjaiah et al. (2015a) observed the hardening effect up to 80 µm depth from the WSEMed surface of $Ti_{50}Ni_{40}Cu_{10}$ SMM. Surface hardness was also affected by the parameters of the WSEM process. An increase of 9.6% in micro-hardness was observed for higher pulse-on-time (128 µs) as compared to lower pulse-on-time (112 µs) and also, approximately 50% increase in hardness was observed with respect to base material. In the WSEM of $Ti_{50}Ni_{40}Co_{10}$ SMM, surface hardness was found to increase by about 30% than the hardness of the bulk material. Hardness near the machined surface (0-100 µm depth) increased due to the formation of various oxides such as TiO_2, Ni and $NiTiO_3$. The hardness of the HAZ also increases; however, the increase is lower than that of the RL, whereas the hardness of the converted layer (beneath HAZ) remains almost the same as the hardness of the bulk material (Soni et al. 2017). Similar hardening effect was also observed during the WSEM of Fe-based SMM. Surface hardness was observed to be 550 Hv for Fe-based SMM, due to the formation of various compounds and solid solutions (Lin et al. 2005). Similar findings have been observed for TiNiX (X = Zr/Cr) ternary SMM (Hsieh et al. 2009), $Ni_{60}Ti_{40}$ SMM (Lotfineyestanak and Daneshmand 2013), $Ni_{50}Ti_{50}$ (Manjaiah et al. 2014) and Nitinol (SE508) (Liu et al. 2014).

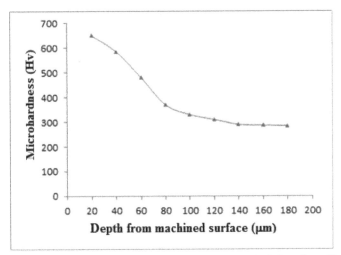

Fig. 2.11: Micro-hardness profile of a machined $Ni_{50.89}Ti_{49.11}$ SMM surface (T_{on} = 125 µs, T_{off} = 54 µs, V_s = 50 V, WF = 8 m/min, WT = 5 N) (Bisaria and Shandilya 2019 © Sage. Reprinted with permission)

2.2.2.4 Shape Recovery Ability (SRA)

As mentioned earlier, the bending test is used to measure the SRA of SMM. When it comes to the WSEM of SMM, it has been noticed that SMM show perfect SRA at normal bending strains; yet, the SRA is slightly compromised at higher bending strains. After WSEM, $Ti_{35.5}Ni_{49.5}Zr_{15}$ and $Ti_{50}Ni_{49.5}Cr_{0.5}$ SMM display perfect SRA at bending strains of 3% and 5%; a slight reduction in their SRA is observed at a higher level of bending strain (8%). The SRA of TiNi-X (X = Cr/Zr) ternary SMM (annealed and WSEMed) at different bending strains has been illustrated in Fig. 2.12. Similarly, for Fe-based SMM, the SRA

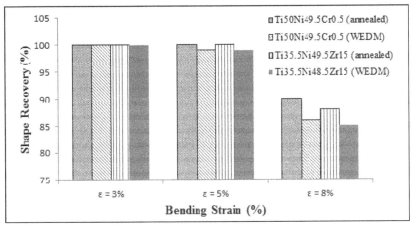

Fig. 2.12: Shape recovery ability of $Ti_{50}Ni_{49.5}Cr_{0.5}$ and $Ti_{35.5}Ni_{49.5}Zr_{15}$ SMM (annealed and WSEMed)

is observed to be lower at a bending strain of 4%, compared to a bending strain of 2% (Lin et al. 2005). The SRA of Nitinol-60 SMM also deteriorates at higher bending strain after WSEM (Lotfineyestanak and Daneshmand 2013).

2.3 Conclusion

This chapter presented a brief introduction of shape memory materials and their machining by WSEM. The machining characteristics and surface integrity aspects of SMM after their machining by WSEM have been described herein. The details presented in this chapter prove that for machining of SMM, the WSEM process is more suitable than any other unconventional machining process. The effects of parameters of the WSEM process on surface integrity aspects such as surface morphology, recast layer, micro-hardness, surface roughness and phase analysis of SMM have also been explored in this chapter.

References

Baumann, M.A. 2004. Nickel-titanium: Options and challenges. Dental Clinics of North America, 48: 55-68.

Bendict, G.F. 1987. Nontraditional Manufacturing Processes. CRC Press, Taylor and Francis. Florida.

Bisaria, H. and Shandilya, P. 2015. Machining of Metal Matrix Composites by EDM and its Variants: A Review. Chapter 23, pp. 267-282. *In:* DAAAM International Scientific Book 2015. B. Katalinic (Ed.). Published by DAAAM International, ISBN 978-3-902734-05-1, Vienna, Austria DOI: 10.2507/daaam.scibook.2015.23.

Bisaria, H. and Shandilya, P. 2018. Experimental studies on electrical discharge wire cutting of Ni-rich NiTi shape memory alloy. Materials and Manufacturing Processes, 33(9): 977-985. DOI: 10.1080/10426914.2017.1388518.

Bisaria, H. and Shandilya, P. 2019. The machining characteristics and surface integrity of Ni-rich NiTi shape memory alloy using wire electric discharge machining. Proceedings of Institution of Mechanical Engineering – Part C: Journal of Mechanical Engineering Science, 233(3): 1068-1078. DOI: 10.1177/0954406218763447.

El-Hofy, H. 2005. Advanced Machining Processes: Nontraditional and Hybrid Machining Processes. McGraw-Hill, New York.

Firstov, G.S., Humbeeck, J.V. and Koval. Y.N. 2006. High temperature shape memory alloys: Problems and prospects. Journal of Intelligent Material Systems and Structures, 17: 1041-1047.

Fujita, A., Fukamichi, K., Gejima, F., Kainuma R. and Ishida, K. 2000. Magnetic properties and large magnetic-field-induced strains in off-stoichiometric Ni-Mn-Al heusler alloys. Applied Physics Letters, 77(19): 3054-3056.

Garg, M.P., Jain, A. and Bhushan, G. 2014. Multi-objective optimization of process parameters in wire electric discharge machining of Ti-6-2-4-2 alloy. Arabian Journal of Science and Engineering, 39: 1465-1476.

Ghosh, P., Rao, A. and Srinivasa, A.R. 2013. Design of multi-state and SMMrt-bias components using shape memory alloy and shape memory polymer composites. Materials and Design, 44: 164-171.

Gil, F.J. and Planell, J.A. 1998. Shape memory alloys for medical applications. Proceedings of the Institution of Mechanical Engineers – Part B: Journal of Engineering Manufacture, 212: 473-488.

Grzesik, W., Kruszynski, B. and Ruszaj. A. 2010. Surface integrity of machined surfaces. pp. 144-158. *In:* Surface Integrity in Machining. Davim, J.P. (Ed.). Springer, New York.

Gupta, K. and Jain, N.K. 2014. On surface integrity of miniature spur gears manufactured by wire electrical discharge machining. The International Journal of Advanced Manufacturing Technology, 72(9-12): 1735-1745. DOI: 10.1007/s00170-014-5772-0.

Hartl, D. and Lagoudas, D. 2007. Aerospace applications of shape memory alloys. Journal of Aerospace Engineering, 221(4): 535.

Hsieh, S.F., Chen, S.L., Lin, H.C., Lin, M.H. and Chiou, S.Y. 2009. The machining characteristics and shape recovery ability of Ti-Ni-X (X=Zr, Cr) ternary shape memory alloys using the wire electro-discharge machining. International Journal of Machine Tools & Manufacture. 49; 509-514.

Huang, W., Ding, Z., Wang, C. Wei, J., Zhao, Y. and Purnawali, H. 2010. Shape memory materials. Materials Today, 13(7-8): 54-61.

Jain, V.K. 2007. Advance machining processes. Allied Publication, New Delhi.

Jameson, E.C. 2001. Electrical Discharge Machining. SME, Dearborn, Michigan.

Jani, J.M., Leary, M., Subic A. and Gibson, M.A. 2014. A review of shape memory alloy research, applications and opportunities. Materials and Design, 56: 1078-1113.

Kamila, S. 2013. Introduction, classification and applications of SMMrt materials: An overview. American Journal of Applied Sciences, 10(8): 876-880.

Kohl, M. 2004. Shape Memory Micro-actuators. Springer, New York.

Kumar, P.K. and Lagoudas, D.C. 2008. Introduction to shape memory alloys. pp. 1-25. *In:* Shape Memory Alloys: Modeling and Engineering Applications. Lagoudas, D.C. (Ed.). Springer, New York.

Kumar, V., Jangra, K.K., Kumar, V. and Sharma, N. 2017. WSEM of nickel based aerospace alloy: Optimization of process parameters and modeling. International Journal on Interactive Design and Manufacturing, 11(4): 917-929. DOI 10.1007/s12008-016-0298-3.

Lin, H.C. and Wu, S.K. 1992. Strengthening effect on shape recovery characteristic of the equiatomic TiNi alloy. Scripta Metallurgica Materialia, 26(1): 59-62.

Lin, H.C., Lin, K.M., Chen, Y.S. and Chu, C.L. 2005. The wire electro-discharge machining characteristics of Fe-30Mn-6Si and Fe-30Mn-6Si-5Cr shape memory alloys. Journal of Materials Processing Technology, 161: 435-439.

Lin, H.C., Lin, K.M. and Chen, Y.C. 2000. A study on the machining characteristics of TiNi shape memory alloys. Journal of Materials Processing Technology, 105: 327-332.

Liu, J.F., Guo, Y.B., Butler, T.M. and Weaver, M.L. 2016. Crystallography, compositions, and properties of white layer by wire electrical discharge machining of nitinol shape memory alloy. Materials and Design. 109; 1-9.

Liu, J.F., Li, L. and Guo, Y.B. 2014. Surface integrity evolution from main cut to finish trim cut in W-EDM of shape memory alloy. Applied Surface Science 308; 253-260.

Lotfineyestanak, A.A. and Daneshmand, S. 2013. The effect of operational cutting parameters on nitinol-60 in wire electrodischarge machining. Advances in Materials Science and Engineering, 2013: 1-6.

Ma, J., Karaman, I. and Noebe, R.D. 2010. High temperature shape memory alloys. International Materials Reviews, 55: 257-315.

Manjaiah, M., Narendranath, S., Basavarajappa, S. and Gaitonde, V.N. 2014. Wire electric discharge machining characteristics of titanium nickel shape memory alloy. Transactions of Nonferrous Materials Society of China, 24: 3201-3209.

Manjaiah, M., Narendranath, S., Basavarajappa, S. and Gaitonde, V.N. 2015a. Effect of electrode material in wire electro discharge machining characteristics of $Ti_{50}Ni_{50-x}Cu_x$ shape memory alloy. Precision Engineering, 41: 68-77.

Manjaiah, M., Narendranath, S. and Basavarajappa, S. 2015b. Wire electro discharge machining performance of TiNiCu shape memory alloy. Silicon, 8: 467-475.

Manjaiah, M., Narendranath, S., Basavarajappa, S. and Gaitonde, V.N. 2015c. Investigation on material removal rate, surface and subsurface characteristics in wire electro discharge machining of $Ti_{50}Ni_{50-x}Cu_x$ shape memory alloy. Proceedings of the Institution of Mechanical Engineers – Part L: Journal of Materials: Design and Applications, 232(2): 164-177. DOI: 10.1177/1464420715619949.

Manjaiah, M., Laubscher, R.F. Narendranath, S., Basavarajappa, S. and Gaitonde, V.N. 2016. Evaluation of wire electro discharge machining characteristics of $Ti_{50}Ni_{50-x}Cu_x$ shape memory alloys. Journal of Materials Research, 31(12): 1801-1808. DOI: 10.1557/jmr.2016.189.

Maruyama, T. and Kubo, H. 2011. Ferrous (Fe-based) shape memory alloys (SMM): Properties, processing and applications. pp. 141-156. In: Shape Memory and Superelastic Alloys, Technologies and Applications. Yamauchi, K., I. Ohkata, K. Tsuchiya and S. Miyazaki (Eds.). Woodhead Publishing, Cornwall, UK.

Mihálcz, I. 2001. Fundamental characteristics and design method for nickel titanium shape memory alloy. Periodica Polytechnica Mechanical Engineering, 45(1): 75-86.

Miyazaki, S. and Sachdeva, R.L. 2009. Shape memory effect and superelasticity in Ti-Ni alloys. pp. 3-18. In: Shape memory alloys for biomedical applications. Yoneyama, T. and T. Miyazaki (Eds.). Woodhead Publishing, CRC Press, Boca Raton, Boston, New York, Washington, DC.

Momoda, L.A. 2004. The future of engineering materials: Multifunction for performance tailored structures. National Academy of Engineering, 34: 1-76.

Morgan, N.B. 2004. Medical shape memory alloy – The market and its products. Materials Science and Engineering: A, 378(1-2): 16-23.

Morgan, N.B. and Broadley, M. 2004. Taking the art out of SMMrt forming processes and durability issues for the application of NiTi shape memory alloys in medical devices. Proceedings of the Materials and Processes for Medical Devices Conference 2004: 247-252.

Manjaiah, M., Narendranath, S., Basavarajappa, S. and Gaitonde, V.N. 2013. Experimental investigations on performance characteristics in wire electro discharge machining of $Ti_{50}Ni_{42.4}Cu_{7.6}$ shape memory alloy. Proceedings of the Institution of Mechanical Engineers – Part B: Journal of Engineering Manufacture. DOI: 10.1177/0954405413478771.

Quan, D. and Hai, X. 2015. Shape memory alloy in various aviation field. Procedia Engineering, 99: 1241-1246.

Rao, A., Srinivasa, A.R. and Reddy, J.N. 2015. Introduction to Shape Memory Alloys: Design of Shape Memory Alloy (SMM) Actuators. Springer, New York.

Sánchez, J.A. and Ortega, N. 2009. Wire electrical discharge machines, machine tools for high performance machining. pp. 307-330. *In:* Machine Tools for High Performance Machining. López de Lacalle, L.N. and A. Lamikiz (Eds.). Springer, London.

Shandilya, P., Jain, P.K. and Jain, N.K. 2012. On wire breakage and microstructure in WEDC of SiCp/6061 aluminum metal matrix composites. The International Journal of Advanced Manufacturing Technology, 61: 1199-1207.

Shimizu, K. 2011. History of the association of shape memory alloys. pp. 197-200. *In:* Shape Memory and Superelastic Alloys, Technologies and Applications. Yamauchi, K., Ohkata, I. Tsuchiya, T. and Miyazaki, K. (Eds.). Woodhead Publishing, Cornwall, UK.

Song, G., Ma, N. and Li, H. 2006. Applications of shape memory alloys in civil structures. Engineering Structures, 28(9): 1266-1274.

Soni, H., Sannayellappa, N. and Rangarasaiah, R.M. 2017. An experimental study of influence of wire electro discharge machining parameters on surface integrity of TiNiCo shape memory alloy. Journal of Materials Research. 32; 3100-3108.

Sun, L. and Huang, W.M. 2009. Nature of the multistage transformation in shape memory alloys upon heating. Metal Science and Heat Treatment, 51(11): 573-578.

Sutapun, B., Tabib-Azar, M. and Huff, M.A. 1998. Applications of shape memory alloys in optics. Applied Optics, 37(28): 6811-6815.

Tadalii, T. 1999. Cu-based shape memory alloys. pp. 97-112. *In:* Shape Memory Materials. Otsuka, K. and C.M. Wayman (Eds.). Cambridge University Press, Cambridge, United Kingdom.

Thompson, S.A. 2000. An overview of nickel–titanium alloys used in dentistry. International Endodontic Journal, 33: 297-310.

Weinert, K., Petzoldt, V. and Kotter, D. 2004. Turning and drilling of NiTi shape memory alloys. CIRP Annals – Manufacturing Technology, 53: 65-68.

Wuttig, M., Li, J. and Craciunescu, C. 2001. A new ferromagnetic shape memory alloy system. Scripta Materialia, 44: 2393-2397.

Spark Erosion Based Manufacturing of Biomedical Components

Sebastian Skoczypiec* and Agnieszka Żyra

Institute of Production Engineering, Cracow University of Technology, al. Jana Pawła II, Kraków, Poland

3.1 Introduction

Due to the worldwide population ageing, expectations towards better life quality and efficient health care systems are increasing. Consequently, demands for customized and sophisticated biomedical devices such as implants, surgical equipment, biomedical instruments, and accessories, devices and systems used during surgeries have been constantly increasing during last several years (Bartolo et al. 2012, Goriainov et al. 2014). The value of the biomedical market, especially medical implants such as joint replacement, spinal and trauma implants, was estimated to be about $30.5 billion in the year 2012 (Goriainov et al. 2014). The statistics presented in Fig. 3.1 depict that the highest growth rate (10.2%) in the global expenditure towards research and development in the biomedical field was observed in 2011; it rose from 0.9% to 4.4% in 2017, and is estimated to be the same during 2018 to 2022. This calls for further improvements in biomedical implants and devices.

The main segment of biomedical industry is the manufacture of biomedical instruments, surgical accessories, devices and equipment, and orthopedic implants from biocompatible materials. Developments in this field are guided by innovations in the design of implants, devices, instruments and equipment, and development of biocompatible materials, rules and regulations. It is important to note that the manufacture of implants, devices, instruments and equipment used for biomedical applications belongs to a very small group of industries. Biocompatibility requirement is a distinguishing feature of the biomedical industry and poses major challenges in selecting from existing materials and manufacturing processes, and the development

*Corresponding author: skoczypiec@mech.pk.edu.pl

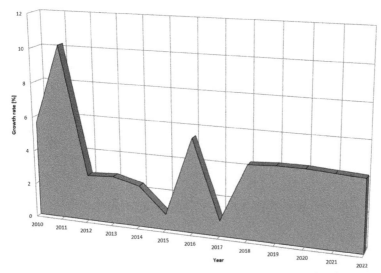

Fig. 3.1: Growth rate of worldwide spending on research and development in the biomedical field (2010-2022)

of new materials and manufacturing processes. Since biocompatibility is defined in the final stage of biomedical component manufacturing, careful selection of the manufacturing technology is thus of utmost importance.

Typical biocompatible materials include austenitic stainless steel, titanium alloys, cobalt-chromium alloys, nickel-based shape memory alloys, and certain types of polymers, ceramics and glasses. Unfortunately, all these materials belong to the category of difficult-to-machine materials; therefore, electro-physical processes are attractive alternatives to manufacture 3D macro and micro features and high aspect-ratio features for biomedical applications in serial and single production of these materials.

In SEM, material removal occurs due to the erosion caused by the occurrence of a series of sparks in the inter-electrode gap (IEG) between the workpiece and the tool electrode, in the presence of a suitably chosen dielectric fluid (Banu and Ali 2016, Jha et al. 2011). Each spark discharge melts and evaporates the material at a single location (Rajurkar et al. 2013). The main function of the dielectric fluid is to flush out the debris produced during the SEM process, act as coolant, and help in solidification of the unflushed material. Many researchers (Kitamura et al. 2013; Kitamura and Kunieda 2014; Koyano et al. 2015; Hayakawa et al. 2016) have tried to investigate the complicated mechanism of material removal during the discharge of a single spark. Occurrence and travel of series of sparks over the entire IEG during short intervals of time makes observation of material removal phenomenon extremely difficult, despite highly sophisticated equipment and gauges. The thermal character of the material removal mechanism causes the formation of the HAZ in the manufactured product. It consists of an upper layer of recast, characterised by a solidified material, micro-cracks, high porosity, larger grains, and varying chemical composition. Additionally, phases other than of the workpiece material can also appear therein, from either the tool electrode or the interaction of the workpiece and the tool materials with the dielectric (Banu and Ali 2016). Hence, achieving a very good surface integrity in SEM is one of the most important challenges faced. This significantly affects the possibilities of using the produced part in a specific environment, especially when a very good surface integrity is needed.

Nevertheless, spark erosion machining (SEM) has the potential to be an effective process for the manufacture of biomedical components such as tools for insertion and extraction (or recovery) of implants, surgical cathodes, syringe components, splints and supports for orthotic and prosthetic devices, bone and jam reamers for dental implants, and tooling and dies for manufacturing and stamping medical equipment and tools. In some cases of implant customization, it can be also used for shaping the surface of surgical screws and bolts for knee, shoulder and hip joint implants. Recent developments in the SEM process have enabled the manufacture of biomedical components from materials with low electrical conductivity and even from isolators (such as ceramics, including glass). However, the main challenge for its application in the biomedical industry is the requirement of biocompatibility. This chapter presents details about the applications of SEM for the manufacture of biomedical components, focusing on biocompatible materials and the issues related to biocompatibility and biotolerance, and their correlation with the applications of SEM.

3.2 Biocompatibility

There are specific requirements in the biomedical industry for each medical device, especially for equipment which are brought in contact with human tissues (like biomedical implants or surgical equipment). Williams (2008) defined biocompatibility of a material as the *ability of a material to perform its desired function with respect to a medical therapy, without eliciting any undesirable local or systemic effects in the recipient or beneficiary of that therapy, but generating the most appropriate beneficial cellular or tissue response in that specific situation, and optimising the clinically relevant performance of that therapy.*

Biocompatibility of a material depends on a following factors: (i) the specific response of an individual to a material, depending on material characteristics as well the situation in which the material is used, and (ii) the specific reaction(s) of a material with the tissues, like degradation over some time in the human body. Biocompatibility of medical equipment depends on the aspects of their manufacture and vice versa. Similarly, manufacture of medicines should also consider biocompatibility.

According to Bartolo et al. (2012), the main pillars of biomanufacturing are bio-design, bio-fabrication, bio-mechatronics and assembly. Currently, the

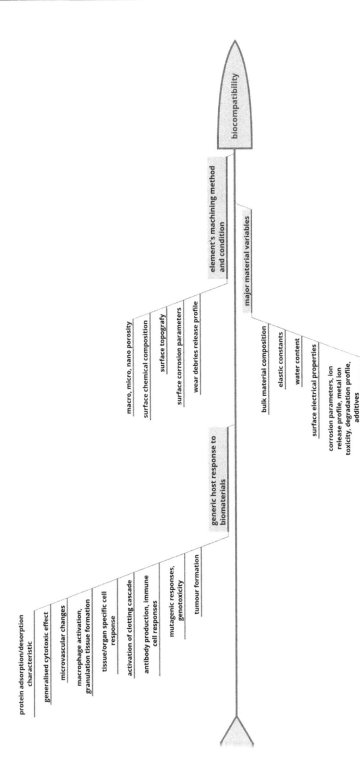

Fig. 3.2: Different factors affecting biocompatibility

most challenging task is to improve the design of implants and shape them in such a way that causes minimal risk of stresses appearing between implants and the surrounding tissues, and consequently reduces the possibility of occurrence of host response and implant degradation (Shayesteh et al. 2016). Figure 3.2 depicts the various factors which affect biocompatibility, underlining the fact that apart from material variables and generic host response to biomaterials, the machining process has considerable influence on the biocompatibility of the medical equipment.

3.3 Biomaterials and their Machinability

The materials used in the medical industry are commonly called biomaterials. It includes all the materials which can act in living organisms, individually or as a parts of complex systems, without interfering with the harmony of interactions in the area of the living matter (Williams 2009). Most commonly used biomaterials can be divided in three groups according to their chemical composition and structure (Bartolo et al. 2012, Shayesteh Moghaddam et al. 2016, Williams 2008): metallic, ceramics and polymers. Metallic biomaterials are used for load-bearing implants or spinal surgical instrumentation due to their good strength, stiffness, toughness, corrosion and impact resistance, mechanical reliability, and biocompatibility (Bartolo et al. 2012, Ramsden et al. 2007, Shayesteh et al. 2016). The second group of biomaterials is ceramics; these have very good compressive strength, inertness, tribological properties, biocompatibility and reliability. They can be used for replacement of hard tissues in living organisms (Bartolo et al. 2012). Some of these ceramics are also bioactive, bioactive glass (used for replacing tissues) for instance. The third group of biomaterials consists of polymers, which are biocompatible, whereas some are biodegradable as well (Bartolo et al. 2012). Properties of polymers can vary extensively on the basis of their applications in the human body. Table 3.1 presents some of the selected biomaterials of each category, along with their applications in the medical industry.

Medical equipment should be machined with special care taken of shape and dimensional accuracy, and superior surface integrity to avoid fatigue or deterioration of implants or surgical devices. For example, spinal instruments must be strong enough to protect and support spine's stability and, simultaneously, ductile enough to prevent stress-shielding, without causing any allergies. Nowadays, the demand for diversified implants and medical/surgical devices is increasing. This requires the capability of designing and manufacturing a variety of shapes, including very small and precise ones, as well as customised implants or sets of surgical instruments from various materials. These materials are characterised by application-dependent mechanical, electrical, thermal and chemical properties. Hence, improvement in the existing machining processes and development of new

Table 3.1: Some of biomaterials (metallic, ceramics and polymers) and their applications in the medical industry (Bartolo et al. 2012, Ramsden et al. 2007, Williams 2008)

Category	Name	Applications in the medical industry
Metallic	Stainless steel	Hip and joint prostheses, stents fracture fixation wires, pins, screws, plates, needles
	Cobalt-based alloys	Bearing surfaces, heart valves, stents, pacemaker leads
	Titanium alloys	Dental implants, femoral stems, pacemaker cans, heart valves, fracture plates, spinal cages, artificial hips and joints, surgical equipment
	Nitinol	Shape memory applications, vascular stents
Ceramics	Alumina	Femoral heads of total hip prostheses, bearing surfaces
	Zirconia	Femoral heads of total hip prostheses, bearing surfaces, shoulder reconstruction, coating over titanium in dental implants
	Bioactive glasses	Tissue replacement
Polymers	Silicones	Soft tissue augmentation, insulating leads, ophthalmological devices
	Poly-tetra-fluoro-ethylene, polyester textile	Vascular grafts, heart valves
	Polyurethane	Pacemaker lead insulation

processes to provide efficient shaping of the tailored medical equipment are necessitated. Most of these can be met by SE-based machining processes.

Since the mechanism of material removal in SE-based machining processes uses electrical and thermal phenomena, the electrical and the thermal properties of the workpiece material determines its machinability. Among electrical properties, electrical resistivity in particular determines the possibility of applying a typical SEM process. Only materials with electrical resistivity below 100 Ωcm (König et al. 1988) can be machined effectively by the non-modified SEM process. However, material composition or technological modification gives the possibility of shaping some difficult-to-machine insulating materials. Mechanical properties of the workpiece material have little effect on machinability and removal efficiency, except for hard and brittle materials like ceramics and metal matrix composites in which spalling plays an important role in material removal. Thermal properties of the workpiece material determine how efficiently the material can be heated up, vaporized and cooled down. Rate of heat absorption and dissipation, which determines the properties of the HAZ, also depend on the thermal properties of the workpiece material. It is worth mentioning that accuracy of the final product and the HAZ composition also rely on the thermal properties of the tool electrode material. Biomaterials encompass a variety of materials with different mechanical, electrical and thermal properties. Table 3.2 lists those properties of biomaterials which affect erosion resistance (that is the machinability of the SEM process).

Table 3.2: Mechanical, thermal and electrical properties of some important biomaterials

Material	Young's modulus [GPa]	Melting point [K]	Electrical resistivity [Ωcm]	Thermal conductivity [W/mK]	Specific heat [J/(kgK)]
Steel (0/15% C)	210	1755	0.000169	54	420
X12CrMoS17 Stainless Steel	200	1673	0.0000600	26.1	460
Ti-6Al-4V alloy	113.8	1877-1933	0.000178	6.70	526.3
Nitinol (martensite)	28 to 41	1573	0.00008	8.6	837
Nitinol (austenite)	83	1573	0.0001	18	837
Cobalt-chromium alloy	210-250	1603	0.009	14.8	452
ZrO_2 ceramics	200-210	2963	> 1e+12	2.20-2.50	420
Pyrex glass (Borosilicate Glass)	64	1073	10^{15}	1.14	753
Silicon	165	1687	–	148	700

Biomaterials can be divided into two groups based on the application of the SEM process:

• Metallic biomaterials, such as stainless steel, titanium alloys, nitinol and cobalt-chromium alloys
• Non-metallic biomaterials, such as ceramics including glasses and polymers which have insulating properties

The application of the SEM process is appropriate for the first group of biomaterials due to technological reasons, as there is no other effective process to give these materials the desired shape. Nevertheless, SEM is one of the processes used for shaping the second group of materials as well, alongside ultrasonic machining (USM), abrasive jet machining (AJM), and abrasive water jet machining (AWJM).

3.4 Spark Erosion Machining of Metallic Biomaterials

Pulse-on time, current, servo-voltage, type of dielectric fluid and tool electrode material used significantly affect the surface integrity of a component manufactured by SEM. Increase in pulse duration reduces the discharge energy density; consequently, MRR decreases, wear of the tool electrode reduces, and the surface quality improves. Increase in current increases the MRR by a single spark discharge; however, it deteriorates the machined surface quality at the same time (Zhang et al. 2016; Torres et al. 2015). Figure 4.3 presents the scanning electron micrographs of surfaces machined by SEM using different currents and dielectrics. It is evident from these images that an increase in the current increases the size of craters and boundary grain growth, regardless of the type of dielectric used; however, use of a carbon-based dielectric increases the number of deep micro-cracks on the surface. Figure 4.3 also depicts that a corrosion resistance test of the SEMed surfaces helps in pit propagation, with its intensity increasing with current. The surfaces machined using a carbon-based dielectric have higher intensity of corrosion in crevices and micro-cracks, compared to those machined using distilled water. Appropriate selection of machining conditions can reduce thermal and structural changes in the surface layer. Żyra et al. (2017) mentioned that the selection of dielectric fluid influences the micro-hardness of the surface layer. For example, out of the surfaces machined by SEM using a carbon-based dielectric and distilled water as the dielectric, the former are harder and more brittle.

One of the disadvantages of SEM is corner, radial and axial wear of the tool electrode. A greater concern here is the diffusion of the melted tool electrode material into the workpiece surface, resulting in significant changes in its chemical composition (Jha et al. 2011, Torres et al. 2015, Risto et al. 2016, Tripathy and Tripathy 2017). Table 3.3 presents the chemical compositions of three materials: 1. the unmachined stainless steel (X5CrNi1810), 2. the SEMed sample using a cylindrical copper tool electrode, distilled water as the dielectric, with a current of 5 A, and 3. the SEMed sample after corrosion resistance test. It can be observed from this table that a certain amount of copper is present in the SEMed sample, indicating that the material melted from the copper electrode was transferred to the surface layer. Also, an increase in the oxygen content after the corrosive resistance test of the SEMed sample indicates that oxidation of the surface layer has also occurred (Żyra et al. 2017).

Zinelis (2007) and Ntasi et al. (2010) investigated the effects of SEM on the corrosion resistance of Co-Cr biatomic alloy (used in dental applications), using kerosene as the dielectric fluid. They concluded that SEM is a good process for dentistry applications; yet, to improve the corrosion resistance of implants, an alternative material should be found to replace the copper electrode. SEMed surfaces have a rough texture with pores, craters and recast layers (thickness about 10 µm); deposition of carbon and copper resulting from the decomposition of the tool electrode and the dielectric can also be found on these surfaces. Table 3.4 presents the secondary electron images and results of the EDX analysis of the surface after conventional finishing and SEM of the Co-Cr alloy before and after corrosion testing. Rough surface texture and change in composition adversely affect the corrosion behaviour

Current = 1 A; Dielectric: Distilled water

Current = 10 A; Dielectric: Distilled water

Current = 1 A; Dielectric: Exsoll D80

Current = 10 A: Dielectric: Exsoll D80

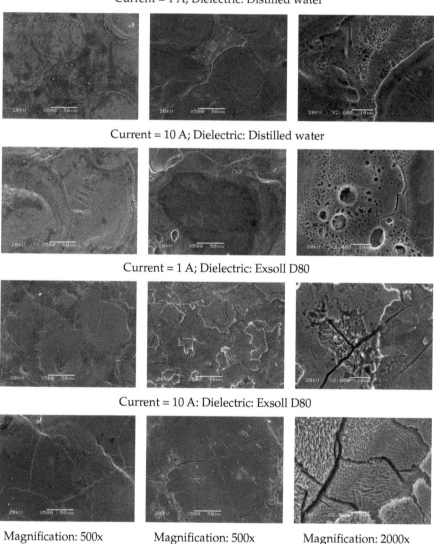

Magnification: 500x   Magnification: 500x   Magnification: 2000x

SEMed suraface   **SEMEd surfaces after corrosion resistance test**

Fig. 3.3: Micrographs of surfaces machined by SEM using distilled water and kerosene as the dielectric fluids (current = 1 A, 10 A)

Table 3.3: Results of the EDX analysis of the stainless steel X5CrNi1810 surface after SEM (current = 5 A; deionised water used as the dielectric)

Element	Unmachined sample [%]	SEMed sample [%]	SEMed sample after its corrosion resistance test [%]
O	2.2	2.4	4.3
Cr	19.3	5.1	4.5
Fe	70.3	78.3	77.4
Ni	7.8	9.9	9.0
Cu	0	4.2	4.5

of the Co-Cr alloy due to galvanic phenomena among different phases and the presence of easily soluble Cu phase. Increased surface roughness also increases susceptibility to pitting and crevice corrosion, which may result in the concentration of highly reactive ions (like chlorine) in areas with pores on the dental implants. Since copper belongs to the group of toxic elements for human body (Klocke et al. 2012), proper and careful choice of tool electrode material, dielectric and other spark erosion process parameters to reduce or eliminate unfavorable surface structures and composition changes is a key issue.

Titanium alloys are one of the popular metallic biomaterials, characterised by relatively higher electrical resistivity and lower thermal conductivity (Table 3.2), compared to stainless steel. Subsequently, an inefficient heat dissipation from the IEG increases the temperature of the SEMed component (Fonda et al. 2008). Furthermore, electrical resistance of titanium alloys increases with temperature; therefore, it is challenging to select proper parameters of the SEM process for machining the titanium alloy. Klocke et al. (2012) have demonstrated the use of WSEM for the manufacture of biomedical components from titanium alloys. Optimization of WSEM parameters and flushing of deionized water can help in attaining an average surface roughness 'R_a' value of less than 0.8 μm. Use of zinc-coated brass wire also aids in minimizing zinc contamination and eliminating toxic copper. Since zinc is very reactive with HCl, post-WSEM treatment can therefore be used for its removal. Experimental results for zinc-coated and uncoated brass wire using optimized WSEM conditions have been summarized in Table 3.5.

One of the recent research trends in implantology is the use of biodegradable implants. Magnesium alloys are one of the most suitable materials for biodegradable orthopedic implants. They have very high biocompatibility and initial structural stability, but their degradation is substantially fast and loss of stability occurs quite early. One of the main reasons affecting the time of implant degradation is the microstructure and the macrostructure of the surface, which should be correctly designed and machined. However, such structures are very difficult to machine by

Table 3.4: Secondary electron images at 1000× magnification and results of EDX analysis of the surface after conventional finishing and SEM of the Co-Cr alloy (before and after corrosion testing) (Ntasi et al. 2010 © Elsevier. Reprinted with permission)

	Conventionally finished surface	Conventionally finished surface after corrosion test	SEMed surface	SEMed surface after corrosion test
Wt.% C	0.9 ± 0.2	5.1 ± 0.5	6.2 ± 0.4	47.1 ± 2.1
Wt.% Cu	Below detection limit	Below detection limit	6.9 ± 0.2	5.5 ± 0.2

Table 3.5: Surface roughness in roughing, finishing and surfacing
of zinc-coated and uncoated wire

Type of wire	Surface roughness Ra [μm]		
	Roughing	Finishing	Surfacing
Zn-coated brass wire	3.45	2.45	0.80
Uncoated brass wire	3.45	2.50	0.85

conventional machining processes, thus making SEM a good alternative. Klocke et al. (2013) studied the influence of sinking SEM and WSEM processes on the biocompatibility of magnesium alloy WE43. They concluded that in the case of rough machining by WSEM, some electrode material is deposited on the machined surface by using CuZn36 wire (Fig. 3.4a) and stainless steel wire (Fig. 3.4b); nevertheless, finish cut can remove these contaminated particles. These problems do not occur with sinking SEM using oil-based dielectric; however, deposition of carbon (released from the dielectric) can be observed on the machined surface. A negative influence of the electrode material particles on the cell viability was confirmed through an *in-vitro* toxicity test; however, the negative influence can be avoided by optimizing the SEM parameters. During the toxicity test, corrosion of the entire magnesium samples occurred. Hence, the use of plasma electrolytic conversion process in the post-processing stage was recommended to create a very high chemical resistance layer on the structured magnesium surface.

Mobile electronic devices such as digital thermometers, pulse oximeters, pulse or blood pressure monitors, weighing scales, glucose meters, cardio exercise machines, electrocardiogram, insulin pumps, etc. constitute a very fast-growing group of biomedical equipment. Most of them are portable and wearable and not invasive, whereas some are implantable (such as a defibrillator). As their market is growing very rapidly, so is the demand for efficient shaping processes for manufacturing these equipment (Konn 2012). The SEM process can be successfully used for machining the materials used in mobile electronic devices for biomedical applications. Their applications include machining of micro-details, micro-electrodes or MEMS parts (Crawford 2012, Heeren et al. 1997), micro-holes (Weng et al. 2006), and silicon cutting (Crawford 2012). Semiconductors can also be machined efficiently by the SEM process, with appropriate doping, using proper polarity according to the type of doping (tool electrode as cathode for p-type semiconductors and as anode for n-type semiconductors), and, in some cases, by metal plating the semiconductor workpiece to improve electrical contact (Heeren et al. 1997).

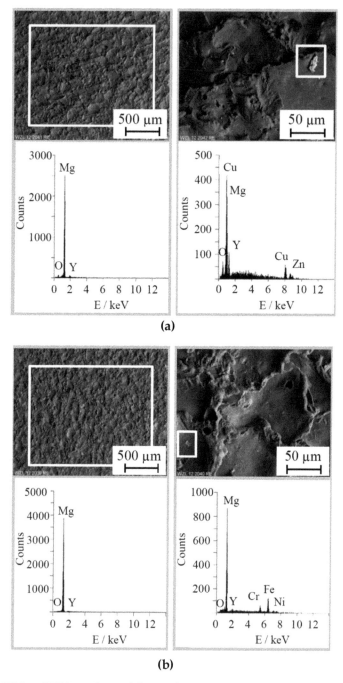

Fig. 3.4: SEM and EDX-analysis of the machined surface of magnesium alloy WE43 after roughing by WSEM using wires of (a) CuZn36 (b) stainless steel (Klocke et al. 2013 © Elsevier. Reprinted with permission)

3.5 Spark Erosion Machining of Non-metallic Biomaterials

Analysis of spark erosion-based machining of non-conductive materials reveals that the development of processes can be classified into the following categories:

- The first category of processes are developed on the principle of SEM (i.e. spark discharges in the presence of a dielectric) by either (i) adding electrically conductive additives (like TiN, carbon nanotubes) to improve the electrical conductivity of the non-conductive biomaterial(s), or (ii) taking help of an electrically conductive coating on the non-conducting workpiece in a process referred to as assisting electrode spark erosion machining (AESEM) (Fig. 3.5a). Since each biomaterial requires certification, adding any conductive phase requires the long and expensive process of material certification to be repeated. For this reason, research on SEM of non-conducting biomaterials such as ZrO_2 or Si_3N_4 focuses on AESEM (Xiaopeng et al. 2012, Hanaoka et al. 2013, Sabur et al. 2013, Ojha 2016, Gotoh et al. 2016).

- Processes in the second category use breakdown of electrolysis in the electrochemical cell (Fig. 3.6) in a process named as spark assisted electrochemical machining (SAEM). Herein, electrochemical reactions between the tool electrode and the auxiliary electrode cause the formation of a hydrogen film around the tool electrode, which serves as a medium wherein the discharge of sparks takes place. It is an emerging micro-machining process especially preferred for the manufacture of micro-channels, micro-grooves, micro-holes and 3D complex shapes in non-conductive materials such as glass, ceramics and quartz. It is worth mentioning that the electrochemical reactions only create favorable conditions for material removal and the material is removed due to the combined mechanism of local heating and chemical etching.

3.5.1 Assisted Electrode Spark Erosion Machining

Engineering ceramics are used in the medical industry as the base material for hip and dental implants, bone scaffolds and fillers in tissue engineering. Traditional manufacturing of ceramic parts consists of the following sequence of processes: powder synthesis, shaping, green machining, de-binding, sintering and final machining. In majority of cases, features of ceramic parts are achieved during their machining in the green or unfired state; however, some final shaping processes are carried out on the sintered state as well (Ferraris et al. 2015). Particularly, such a need arises while shaping complex 3D parts, micro-parts or small volume production. SEM can be directly applied to a fully dense and hard ceramic blank. Since most engineering ceramics such as Al_2O_3, ZrO_2, Si_3N_4 and SiC are electrically non-conducting, the use of SEM for their machining has led to the development of assisting electrode SEM (AESEM).

In AESEM, a layer of conductive material is coated over the non-conductive workpiece material initially. The role of this layer is to allow the discharge of sparks at the start of the SEM process. A thin conductive layer of carbon, which comes from the hydrocarbon based dielectric fluid, is created on the machined surface during the SEM process because of the discharge of series of sparks. This is referred to as the in-process conductive layer (Fig. 3.5). The parameters of the AESEM process should be selected in such a way that allows material removal and in-process formation of a conductive layer of carbon. The nature of the process and the properties of ceramics (like high melting temperature, high hardness and brittleness) play an important role in melting and vaporization during material removal in the AESEM process, causing spalling and, in some cases, thermal decomposition (thermolysis) and carbide formation as well.

Reliability and efficiency of AESEM depends mainly on proper preparation of the initial conducting layer and the stability of the in-process developed carbon layer, which in turn depend on the selection of appropriate AESEM parameters. The following solutions are possible for the initial conducting layer (Ojha 2016):

- Coating of a thin layer of metallic materials such as copper or aluminum, or applying a mesh without additional adhesive on the ceramic surface (the time required to form the in-process carbon layer is high and the machining process is not stable)
- Deposition of a metallic layer by chemical vapor deposition (CVD) or physical vapor deposition (PVD) (the thickness of the initial layer should be more than 10 µm for stable SEM, but achieving such thickness by CVD and PVD is expensive and not efficient)
- Applying a carbon-based lacquer for smooth transition from initial to in-process conductive layer; this is economical

The stability of ceramics in AESEM depends significantly on the stability of the in-process developed conductive layer. It mainly consists of carbon and constituents of the materials of the tool electrode; its thickness can reach up to 50 µm. This layer remains on the surface after the AESEM process; however, it can be easily removed through heating in the post-processing stage. Machinability of ceramics by AESEM depends primarily on the free energy of formation of carbides (ZrC). AESEM can be successfully applied to structure ceramics such as Si_3N_4, SiC and ZrO_2 as these materials are relatively low energy materials. Free energy of forming Al_4C_3 is much higher; therefore, process stability problems arise while machining Al_2O_3 by AESEM. Hence, only low purity Al_2O_3 can be machined by AESEM easily, machining of pure Al_2O_3 is impossible by this method.

Xiaopeng et al. (2012) presented an interesting variant of the AESEM process referred to as double electrode synchronous servo electrical discharge grinding (DESS-EDG). It uses an additional sheet metal electrode which is automatically fed between the conductive grinding wheel and the workpiece (Fig. 3.5b). Discharge of sparks occurs between the grinding wheel and the thin (0.05 ~ 0.15 mm) sheet electrode to remove material from the non-conductive workpiece. Additionally, the grinding wheel removes the SEMed layer. Combination of SEM with grinding in this process gives the possibility to machine non-conductive ceramics with 2 to 4 times higher efficiency and significantly lower forces compared with the simple grinding process. This process, similar to electro discharge grinding (EDG), requires additional effort for dressing the grinding wheel to maintain its shape and accuracy.

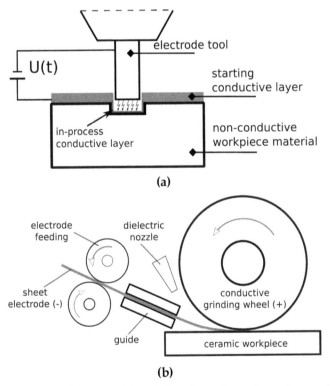

Fig. 3.5: Schematic diagrams: (a) assisting electrode spark erosion machining (AESEM) (b) double electrode synchronous servo spark erosion grinding process

3.5.2 Spark-assisted Electrochemical Machining

The phenomenon of electrochemical discharge (also called electrode effect or plasma electrolysis) is used to remove material from the non-conductive workpiece in spark assisted electrochemical machining process (SAEM). An auxiliary or counter electrode is used as the anode in this process (Fig. 3.6)

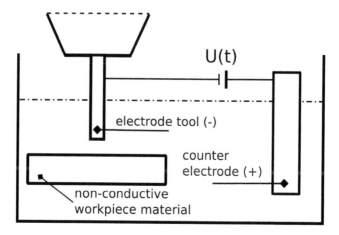

Fig. 3.6: Schematic of spark assisted electrochemical machining (SAEM)

The auxiliary electrode has a surface area significantly larger (at least 100 times higher) than that of the tool electrode and the distance between these electrodes is in range of centimetres. Typically, continuous or pulsed DC voltage in the range of 30-40 V is applied between the cathodic tool electrode and the anodic counter electrode (Cheng et al. 2010). The electrochemical reactions between the tool electrode and the counter electrode immersed in an electrolytic solution cause the formation of a hydrogen gas film around the tool electrode, thus isolating it. When the applied voltage is increased beyond the critical value, breakdown of the electrolyte is observed. The electrical field in this area is high enough to allow the discharge of sparks between the tool electrode and the electrolyte. Since the workpiece is usually placed near the tool electrode (< 1 mm), an unwanted effect therefore removes the material from the non-conducting workpiece. The value of critical voltage significantly depends on the geometry of the electrochemical cell (surface area and the distance between the electrodes), and the electrolyte concentration. (Wu 2005, Wüthrich and Mandin 2009, Wüthrich and Allagui 2010, Jiang et al. 2014). Following are the typical events that occur during the SAEM process:

- Melting and evaporation
- High temperature dissolution
- Thermal stresses, micro-cracking and spalling
- Mechanical impact of expanding gases and moving electrolyte
- Electrochemical etching

Stability and continuity of material removal result from an appropriate feeding mechanism of the tool electrode. In majority of the cases, gravity feed is used, which enables a constant force between the tool electrode and the workpiece, or a constant feed rate. Application of SAEM is justified for materials difficult-to-machine by nontraditional processes such as EDM, ECM or LBM. Non-conducting biomaterials such as pyrex glass and Al_2O_3 belong to such category. With SAEM, the workpiece can be shaped by drilling, milling or cutting. The size of the machined shape does not exceed a few millimetres. For example, it is in the range of 0.1 to 2 mm for glass, and between 0.1 and 1 mm for Al_2O_3 ceramic.

3.6 Plasma Electrolytic Polishing

Plasma electrolytic polishing (PEP) is a surface finishing process used to obtain a very smooth and high-gloss surface on metals and alloys (Nestler et al. 2016, Zeidler et al. 2016). In this process, the workpiece is immersed into an electrolyte solution of low conductivity (80-140 mS/cm) and is specific to the anodic workpiece material. The cathodic electrode tool is also immersed in the same tank. Application of a DC voltage in the range of 180 to 400 V results in the electrolytic dissolution of the workpiece material, oxide and hydrogen formation and alkalization, formation of plasma skin (ionization) over the workpiece, and hydrothermal reactions. According to Nestler et al. (2016), the smallest possible surface roughness obtainable with this method is about 0.02 μm; however, it significantly depends on the initial value of surface roughness (before PEP). Extremely rough surfaces are difficult and sometimes even almost impossible to finish. Moreover, the material removal rate is also low (less than 5 μm/min). Increase in process efficiency leads to a loss in the surface gloss effect. The PEP process can finish complicated sculptured surfaces without the need of a shaped electrode tool. Moreover, it does not alter the chemical composition of the workpiece material. This aspect is very important in the manufacture of components and devices for biomedical applications. However, the following disadvantages limit the applications of PEP: (i) need for a very careful selection of electrolyte for each material, (ii) one-step polishing is not practically possible, (iii) significant decrease in edge sharpness with long processing, and (iv) problems with finishing of small inner cavities.

Zeidler et al. (2016) showed that PEP does not yield any adverse effect on the biocompatibility of components and on cell life. *In-vivo* test of PEP-finished dental-corona showed improvements in resistance to bacterial growth, thus limiting plaque. This shows the potential and promise of PEP for the finishing of biomedical components. However, further development and maturity of PEP demands sustained research on the issues of biocompatibility, and an increase in the material application range. One prominent application of PEP in the medical industry includes the polishing of surgical instruments, polishing of orthopedic and dental implants made of Co-Cr and titanium, and finishing of additively manufactured biomedical parts as shown in Fig. 3.7.

(a)

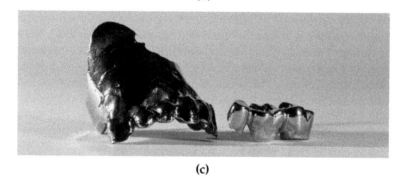

(b)

(c)

Fig. 3.7: Additive manufactured orthopedic and dental implants finished by plasma electrolytic polishing (Source: https://beckmann-institut.de)

3.7 Summary

Application of spark-based manufacturing processes in the medical industry results from their ability to machine the biocompatible materials, regardless of their mechanical properties and chemical structure. Some modifications in the SEM process allow the machining of non-conducting biomaterials as well. The second aspect of concern is the possibility of shaping complex shapes (like small holes, precision tapers, thin walls and webs, cavities with sharp corners, and parts with high-aspect ratio) with relatively high accuracy and achievable roughness. Furthermore, the SEM process is effective for small production run and single parts; therefore, customized implants, tooling and prototypes are important areas of further research. This is very important, especially keeping in view the dynamic development of reconstruction medicines where the manufacture of tailor-made components becomes essential more often.

According to specific applications, SEM is used directly for machining biomedical and dental components, implants and tools (like precision stainless-steel guides for bone saws), as well as indirectly in the manufacture of tools, dies and molds (like mold cavities for injection-molded plastic syringes) during serial or prototyping production. Features of SEM process also correspond with the increasing demand for biomedical device and surgical equipment miniaturization. Greater emphasis on the reduction of surgical area during surgeries necessitates smooth, precisely-shaped surfaces and edges of surgical tools used for precise tissue removal. This makes ultraprecise micromachining technologies increasingly essential. Manufacturing of microfluidics devices and sensors, where machining precision determines the functioning in humans, is another emerging area.

Possibilities of applications of SE-based machining are related with the biological properties and corrosion resistance of the surface obtained in the process. Changes in the chemical composition of biomaterials and biocompatible materials are also a crucial factor here. Therefore, further research to optimize the machining parameters and to characterize the biological responses of these materials is required.

References

Banu, A. and Ali, M.Y. 2016. Electrical discharge machining (EDM): A review. International Journal of Engineering Materials and Manufacture, 1(1): 3-10, DOI: 10.26776/ijemm.01.01.2016.02

Bartolo, P., Kruth, J.P., Silva, J., Levy, G. and Malshe, A. 2012. Biomedical production of implants by additive electro-chemical and physical processes. CIRP Annals: Manufacturing Technology, 61(2): 635-655.

Cheng, C.P., Wu, K.L., Mai, C.C., Yang, C.K., Hsu, Y.S. and Yan, B.H. 2010. Study of gas film quality in electrochemical discharge machining. International Journal of Machine Tools and Manufacture, 50(8): 689-697.

Crawford, G.A. 2012. Process characterization of electrical discharge machining of highly doped silicon. http://hdl.handle.net/1721.1/74893

Ferraris, E., Vleugels, J., Guo, Y., Bourell, D., Pierre, J. and Lauwers, B. 2015. Shaping of engineering ceramics by electrochemical and physical processes. CIRP Annals: Manufacturing Technology, 65(2): 761-784.

Fonda, P., Wang, Z., Yamazaki, K. and Akutsu, Y. 2008. A fundamental study on Ti-6Al-4V thermal and electrical properties and their relation to EDM productivity. Journal of Materials Processing Technology, 202(1-3): 583-589.

Goriainov, V., Cook, R., Latham, J.M., Dunlop, D.G. and Oreffo, R.O.C. 2014. Bone and metal: An orthopedic perspective on osseo-integration of metals. Acta Biomaterials, 10(10): 4043-4057.

Gotoh, H., Tani, T. and Mohri, N. 2016. EDM of insulating ceramics by electrical conductive surface layer control. Procedia CIRP, 42: 201-205.

Hanaoka, D., Fukuzawa, Y., Ramirez, C., Miranzo, P., Osendi, M.I. and Belmonte, M. 2013. Electrical discharge machining of ceramic/carbon nanostructure composites. Procedia CIRP, 6: 95-100.

Hayakawa, S., Kusafuka, Y., Itoigawa, F. and Nakamura, T. 2016. Observation of material removal from discharge spot in electrical discharge machining. Procedia CIRP, 42: 12-17.

Heeren, P.H., Reynaerts, D., Van Brussel, H., Beuret, C., Larsson, O. and Bertholds, A. 1997. Microstructuring of silicon by electro-discharge machining (EDM) – Part II: Applications, Sensors and Actuators, 61: 379-386.

Jha, B., Ram, K. and Rao, M. 2011. An overview of technology and research in electrode design and manufacturing in sinking electrical discharge machining. Journal of Engineering Science Technology Review, 4(2): 118-130.

Jiang, B., Lan, S., Ni, J. and Zhang, Z. 2014. Experimental investigation of spark generation in electrochemical discharge machining of non-conducting materials. Journal of Materials Processing Technology, 214(4): 892-898.

Kitamura, T. and Kunieda, M. 2014. Clarification of EDM gap phenomena using transparent electrodes. CIRP Annals: Manufacturing Technology, 63(1): 213-216.

Kitamura, T., Kunieda, M. and Abe, K. 2013. High-speed imaging of EDM gap phenomena using transparent electrodes. Procedia CIRP, 6: 314-319.

Klocke, F., Welling, D., Dieckmann, J. and Klink, A. 2012. Titanium parts for medical sector made by Wire-EDM. Proceedings of 1st International Conference on Design and Processes for Medical Devices (PROMED), Brescia, Italy, pp. 164-167.

Klocke, F., Schwade, M., Klink, A., Veselovac, D. and Kopp, A. 2013. Influence of electro discharge machining of biodegradable magnesium on the biocompatibility. Procedia CIRP, 5: 88-93.

König, W., Dauw, D.F., Levy, G. and Panten, U. 1988. EDM – Future steps towards the machining of ceramics. CIRP Annals: Manufacturing Technology, 37(2): 623-631.

Konn, J. 2012. Overview of the medical semiconductor market and applications, http://medsmagazine.com/2012/04/overview-of-the-medical-semiconductor-market-and-applications/, accessed on 10.11.2017

Koyano, T., Hosokawa, A., Suzuki, S. and Ueda, T. 2015. Influence of external hydrostatic pressure on machining characteristics of electrical discharge machining. CIRP Annals: Manufacturing Technology, 64(1): 229-232.

Nestler, K., Bottger-Hiller, F., Adamitzki, W., Glowa, G., Zeidler, H. and Schubert, A. 2016. Plasma electrolytic polishing: An overview of applied technologies and current challenges to extend the polishable material range. Procedia CIRP, 42: 503-507.

Niinomi, M., Narushima, T., Nakai, M. and Reactions, B. 2015. Advances in Metallic Biomaterials: Tissues, Materials and Biological Reactions, Springer-Verlag Berlin Heidelberg.

Ntasi, A., Dieter, W., Eliades, G. and Zinelis, S. 2010. The effect of electro discharge machining (EDM) on the corrosion resistance of dental alloys. Dental Materials, 26(12): 237-245.

Ojha, N. 2016. Electrical Discharge Machining of Non-Conductive Advanced Ceramics. PhD Thesis, Albert-Ludwigs-Universität Freiburg.

Rajurkar, K.P., Sundaram, M.M. and Malshe, A.P. 2013. Review of electrochemical and electrodischarge machining. Procedia CIRP, 6: 13-26.

Ramsden, J.J., Allen, D.M., Stephenson, D.J., Alcock, J.R. and Peggs, G.N. 2007. The design and manufacture of biomedical surfaces. CIRP Annals: Manufacturing Technology, 56(2): 687-711.

Risto, M., Haas, R. and Munz, M. 2016. Optimization of the EDM drilling process to increase the productivity and geometrical accuracy. Procedia CIRP, 42: 537-542.

Sabur, A., Ali, M.Y., Maleque, M.A. and Khan, A.A. 2013. Investigation of material removal characteristics in EDM of nonconductive ZrO_2 ceramic. Procedia Engineering, 56: 696-701.

Shayesteh, M., Taheri Andani, M.N., Amerinatanzi, A., Haberland, C. and Huff, S. 2016. Metals for bone implants: Safety, design, and efficacy. Biomanufacturing Review, 1(1): 1.

Torres, A., Puertas, I. and Luis, C.J. 2015. Modelling of surface finish, electrode wear and material removal rate in electrical discharge machining of hard-to-machine alloys. Precision Engineering, 40: 33-45.

Tripathy, S. and Tripathy, D.K. 2017. Optimization of process parameters and investigation on surface characteristics during EDM and powder mixed EDM. pp. 385-391. *In:* Innovative Design and Development Practices in Aerospace and Automotive Engineering: I-DAD, 22-24 February. R.P. Bajpai and U. Chandrasekhar (Eds.). Singapore: Springer Singapore.

Weng, F.T., Hsu, C.S. and Lin, W.F. 2006. Fabrication of micro components to silicon wafer using EDM process. Material Science Forum, 505-507: 217-222.

Williams, D.F. 2008. On the mechanisms of biocompatibility. Biomaterials, 29(20): 2941-2953.

Williams, D.F. 2009. On the nature of biomaterials. Biomaterials, 30(30): 5897-5909.

Wüthrich, R. and Allagui, A. 2010. Building micro and nanosystems with electrochemical discharges. Electrochimica Acta, 55(27): 8189-8196.

Wüthrich, R. and Mandin, P. 2009. Electrochemical discharges: Discovery and early applications. Electrochimica Acta, 54(16): 4031-4035.

Wüthrich, R. and Fascio, V. 2005. Machining of non-conducting materials using electrochemical discharge phenomenon: An overview. International Journal of Machine Tools and Manufacture, 45(9): 1095-1108.

Xiaopeng, L., Yonghong, L. and Renjie J. 2012. A new method for electrical discharge machining of non-conductive engineering ceramics. Proceedings of 2nd International Conference on Electronic and Mechanical Engineering and Information Technology (EMEIT), pp. 1266-1269. DOI: 10.2991/emeit.2012.280

Zeidler, H., Hiller, B.F., Edelmann, J. and Schubert, A. 2016. Surface finish machining of medical parts using plasma electrolytic polishing. Procedia CIRP, 49: 83-87.

Zhang, M., Zhang, Q., Zhu, G., Liu, Q. and Zhang, J. 2016. Effects of some process parameters on the impulse force in single pulsed EDM. Procedia CIRP, 42: 627-631.

Zinelis, S. 2007. Surface and elemental alterations of dental alloys induced by electro discharge machining (EDM). Dental Materials, 23(5): 601-607.

Żyra, A., Bogucki, R. and Skoczypiec, S. 2017. Austenitic steel surface integrity after EDM in different dielectric liquids. Technical Transactions, 12: 231-242, DOI: 10.4467/2353737XCT.17.222.7765 (https://beckmann-institut.de (accessed on 02.11.2017)

Spark Erosion Machining of Aerospace Materials

Asma Perveen[1]* and Samad Nadimi Bavil Oliaei[2]

[1] Nazarbayev University, Astana, Kazakhstan
[2] Atilim University, Ankara, Turkey

4.1 Aerospace Materials and their Applications

In the last century, the aerospace industry was completely dominated by aluminium alloys due to their better strength-to-weight ratio and relative inexpensiveness. As much as 70% of aeroplane parts were made of aluminium alloys, whereas the use of other materials such as titanium alloys, graphite, and composites was limited to merely 3-7%. However, the trend of materials used for aeroplane parts has changed tremendously with the passage of time and different materials have been introduced for aerospace applications. Less critical aeroplane parts started using lighter-weight composite and honeycomb materials, while critical parts (such as the plane engine) came to use alloys capable of resisting high temperatures. Maximum possible weight reduction is the ultimate goal for every manufacturer of aeroplanes so as to reduce running expenses and increase fuel efficiency. Since the margin for errors for an aeroplane cruising at 35,000 ft altitude is negligible, therefore aerospace industries demand a very high level of accuracy and precision for each part used in the manufacture of an aeroplane, compared to any other industry.

With increasing demands for a lighter, faster and cheaper aeroplane, manufacturers of aeroplanes and aeroplane parts have been exploring every possible new aerospace material. Titanium alloys, nickel alloys, composites, and ceramics are the most widely tried and exploited materials for various aerospace applications. Although the use of aluminium alloys has reduced in aeroplane manufacturing, aluminium is still being used as an alloying element with other metals, such as in titanium alloy Ti-6Al-4V. Another such aluminium-containing alloy is Aluminium 2014-T6, with extensive

*Corresponding author: asma.perveen@nu.edu.kz

applications in the manufacture of structural components (such as tension members and shear webs/ribs) of aeroplanes. The front part of the aeroplane turbine engine, the fan, and low/high pressure compressors are mostly made of titanium alloys, whereas the hot combustion chamber of the turbine engine, the turbine blades, the turbine exhaust case, and low/high pressure turbine sections need a material that can withstand high temperature, like nickel-based alloys. General Electric pioneered the use of Ti-Al alloy to manufacture low-pressure turbine blades in their GEnx engine. Subsequently, even Boeing started using this alloy for manufacturing the jet engine for their 787 Dreamliner. Metal matrix composites (MMC), aluminium-based MMC in particular, have a huge potential in the aerospace industry due to their high strength and light weight.

Machining of titanium alloys is extremely difficult due to their low thermal conductivity and chemical affinity. High speed machining of these alloys using conventional machining suffers from extremely high tool wear and poor surface integrity due to strain hardening, pseudo-elastic behaviour and high toughness. Similarly, nickel alloys have limitations such as the presence of hard abrasive particles in their microstructure, poor machinability and thermal affinity, and work hardening. Aluminium-based MMCs have become popular due to their high temperature retention properties. However, they belong to the category of difficult-to-machine materials due to the presence of hard ceramic particles and poor machinability using conventional machining processes.

Non-conventional machining processes, especially thermal-type processes such as spark erosion machining (SEM), laser beam machining (LBM), plasma arc machining (PAM), and electron beam machining (EBM), have drawn profound attention to overcome the difficulties encountered during traditional machining of aerospace materials. In SEM, a series of spark occurs between the cathode and the anode, in the presence of an appropriate dielectric, once the electric field becomes strong enough to break down the dielectric through ionization, thus making it electrically conducting. Consequently, material from both the electrodes is removed, thus forming small craters on the surface. Being a non-contact machining process, SEM has the advantage of experiencing negligible cutting force, mechanical stress, machining chatters and vibrations, although tool wear is significant. This chapter presents details of SEM of aerospace materials.

4.1.1 Aluminium Alloys

The aerospace industry is in a constant search for new materials or any plausible new concepts that can improve the efficiency of air transportation and reduce the cost (Heinz et al. 2000). Aluminium and its alloys are reasonable options for aeroplane manufacturers due to the low density (almost one-third of that of steel) and better corrosion resistance of aluminium. Reducing the weight of an aeroplane by two-third significantly reduces fuel usage, while increasing its weight carrying capacity. Furthermore, when an aeroplane is subjected to maritime climate, corrosion resistance of its external components becomes important. Other than external parts like fuselage or wings, around 70% of the internal structure of modern aeroplanes like Boeing 777 or A330 is made of Al alloys (Fig. 4.1a). While being economical and lightweight, aluminium alloys also offer reasonably high strength as well as ease of manufacturing (Prasad and Wanhill 2017). Aluminium alloys have sufficient strength (and consequently higher strength-to-weight ratio) for making load-bearing components of an aeroplane, thus replacing heavy metallic parts, without compromising the strength. However, steel is preferred over aluminium for those components of an aeroplane which need much higher strengths, that is, the landing gear. Al2024, is known for its high strength-to-weight ratio and is used in wings or fuselage. Al6061 has similar applications. Al5052 is mostly used in maritime climates due to its excellent corrosion resistance. Al7050/7075 is considered as the best option for aerospace due to its fracture resistance and corrosion resistance. Al7075 was extensively used in Mitsubishi A6M Zero fighter planes during World War II. Al7068, known as one of the toughest alloys, is lightweight and has excellent corrosion resistance. Apart from these alloys, aluminium-based MMCs have also drawn considerable attention because of their high specific strength, stiffness, low thermal expansion coefficient, and dimensional stability at enhanced temperatures. Examples of such AMMC include Al/ SiC, Al/Al$_2$O$_3$ and Al.Ti (Kandpal et al. 2015).

4.1.2 Titanium Alloys

Density of titanium is slightly more than half of that of steel. Titanium alloys provide a higher specific strength and excellent corrosion resistance. This makes titanium alloys a very strong alternative to aluminium alloys for the aerospace industry. Despite titanium alloys being expensive, aeroplane industries still use approximately 10 wt.% titanium alloys for manufacturing both critical and non-critical components of an aeroplane (such as wings, fuselage, frames, tubing for hydraulic lines, fasteners, engine parts, rivets, undercarriage parts, fan-type gas turbine engine, and stator-rotors blades of a compressor). Though replacement of aluminium alloys by titanium alloys may increase the aeroplane weight by about a third (Peters and Leyens 2009), titanium alloys reduce aeroplane weight compared to steel or nickel alloys. Moreover, they provide much better corrosion resistance and temperature resistance than aluminium alloys and steel. Their application as narrow bands surrounding the fuselage prevents any crack from propagating on the outer skin of the fuselage. These alloys are also used in the hydraulic tube of aeroplanes. Aeroplane floor surrounding the kitchen and toilet also requires high corrosion resistance, but with lower strength, and pure titanium serves as good a candidate for such applications (Peters et al. 2003). Use of titanium alloys in the aeroplane industry has been continuously growing over the last century (Fig. 4.1b). Use of different materials in gas turbines for aerospace applications has been shown in Fig. 4.1c. Manufacturing components and products from titanium alloys using conventional processes is energy consuming and expensive and may damage their useful properties. SEM process in this context plays a very important role.

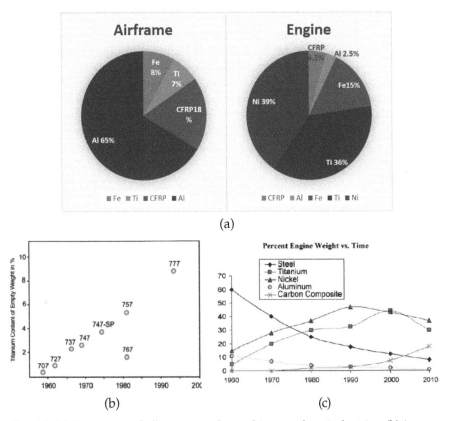

Fig. 4.1: (a) Percentage of alloys currently used in aeroplane industries; (b) increase in the use of titanium alloys in different commercial aeroplanes of Boeing (Peters et al. 2003, © John Wiley and Sons. Reprinted with permission); (c) Use of different materials in gas turbines for aerospace applications (Fonda et al. 2008, © Elsevier Inc. Reprinted with permission)

4.1.3 Nickel Alloys

Nickel-based alloys offer higher toughness, good electrical conductivity, and higher resistance to temperature, corrosion and wear. Consequently, they are quite suitable for use in the aerospace industry. Civilian and military jet engines currently use Nimonic alloy 75 and Inconel alloys. Inconel alloy 600 is widely used in jet engines and airframe components such as exhaust liners, lock wires, and turbine seals due to its higher strength and resistance to corrosion and elevated temperatures. Inconel 601 alloy finds applications in jet engines, aerospace and marine engines due to its higher mechanical strength, and resistance to elevated temperatures, oxidation and corrosion. Good machinability, formability and weldability of this alloy make it one of the most widely used alloys of nickel. Inconel 713CC is used in experimental jet engines as well as turbines, especially in vanes and first stage blades of a jet engine, due to its higher strength, stability and ductility at raised temperatures. Nimonic alloy 75 is used in the form of sheet metals and provides resistance to oxidation scaling at higher operating temperatures. UDIMET 720 is another important superalloy used in aerospace applications such as discs and blades of commercial engines. Owing to its higher strength, and resistance to elevated temperatures and oxidation, Haste alloy is used for manufacturing combustion chambers, afterburners and tail pipe of aeroplanes. Land-based gas turbine engines also use this alloy (Ezugwu et al. 1999). Figure 4.2 shows different parts of a typical aero-engine and different alloys used for manufacturing them.

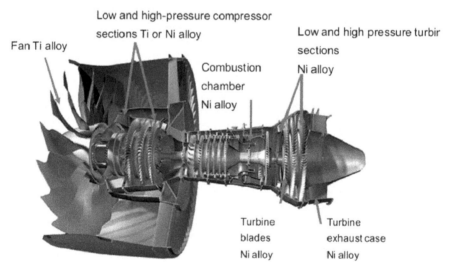

Fig. 4.2: Different components of a typical aero-engine and different alloys of titanium and nickel used in their manufacture (Ulutan and Ozel 2011, © Elsevier. Reprinted with permission)

4.2 Spark Erosion Machining of Aerospace Alloys

Spark erosion machining is one of the most crucial non-conventional machining processes, the performance of which does not depend on the mechanical properties of the workpiece material. However, electrical conductivity of the workpiece and the tool material is a necessity and their thermal properties are even more relevant. Recent developments in assistive machining can be used for SEM of semi-conductors and non-conducting materials as well. Since there is no contact between the tool and the workpiece, no mechanical stress on the workpiece material, chatter, and vibration are generated (Kalpajian and Schmid 2003). The SEM process involves the conversion of electrical energy into thermal energy through a series of electrical sparks between the tool and the workpiece electrodes when the applied DC voltage is enough to break down the dielectric by ionizing it (thus making it electrically conducting). This generates very high temperatures, which can be as high as 20,000°C, for a fraction of a second, consequently melting and vaporizing the materials from both the electrodes. As soon as the voltage supply ceases, the plasma channel breaks down, creating a vacuum to draw fresh dielectric which flushes out the removed material from the machining zone and prepares it for the next cycle of sparks. The volume of the material removed is around 10^{-4} to 10^{-6} mm^3 per spark. Shape, and dimensional and geometrical accuracy of the tool electrode determine the dimensional and geometrical accuracy of the surface generated by SEM (Jameson 1983, McGeough 1988, Boothroyd and Knight 1989, Krar and Check 1997, Tsai et al. 2003). The material removal mechanism involves diffusion of material from one electrode to the other electrode, in solid, molten, or gaseous state, while also causing the alloying reaction with the electrodes' surface. The state of the diffused material, the polarity of the electrodes, and pulse duration influence the material removal phenomenon. Use of short pulses causes most of the material removal due to the electrostatic force and stress distribution on the cathode. Thermal spalling phenomenon is also observed in SEM of ceramic composites due to the existence of an abrupt temperature gradient (Roethel et al. 1976, Erden 1983, Gadalla and Tsai 1989, Lee and Lau 1991, Gangadhar et al. 1992, Singh and Ghosh 1999).

The parameters involved in SEM process can be divided into electrical and non-electrical parameters. Electrical parameters include supply voltage, peak current, polarity, pulse-on time or pulse duration, and pulse-off time. Non-electrical parameters are tool rotation, type, flow rate, pressure and flushing mechanism of the dielectric, and type and shape of the tool material. Investigations have been conducted on SEM and wire SEM (WSEM) of aerospace alloys for studying the effects of SEM process parameters on various performance measures such as material removal rate (MRR), surface integrity, and tool wear. Following sections provide more insight into these aspects.

4.2.1 Material Removal Rate (MRR)

Material removal rate is the ratio of the total quantity of a material removed to the total time taken for doing so. It can be expressed on volumetric and gravimetric bases. It is one of the most prominent performance parameters for any machining process as it indicates its productivity. Literature reports that MRR is significantly influenced by discharge current and duty factor (the ratio of pulse-on time to the sum of pulse-on and pulse off times).

MRR increases with an increase in discharge current, which strengthens the plasma channel. It causes melting and removal of more material easily. MRR also increases with increase in pulse-on time, up to a certain value, and starts decreasing thereafter, thus indicating the existence of an optimum value of pulse-on time. This phenomenon can be explained by the fact that increasing the pulse-on time beyond its optimum value exposes the workpiece surface to the plasma channel for longer durations, thus causing increased local temperatures due to less flushing (pulse-off) time. This results in the decomposition of the dielectric and the bonding of the decomposed material with the workpiece surface. As a result, discharge energy and, in turn, MRR decrease. Positive polarity of the tool electrode increases the MRR due to the higher momentum of positively charged ions striking the workpiece surface (Manjaiah et al. 2014).

Discharge current, pulse-on time and pulse frequency (reciprocal of sum of pulse-on and pulse-off times) significantly affect the MRR in WSEM (Scott et al. 1991). Liao et al. (1997), using Taguchi approach, and Tsai et al. (2003), using grey rational and signal to noise (S/N) ratio, observed that MRR is considerably affected by wire feed rate and pulse-on time. MRR increases with pulse energy, a function of voltage and capacitance. When the voltage is higher, spark discharge can occur with even a bigger inter-electrode gap (IEG). Capacitance determines pulse frequency, which in turn influences crater size (Singh et al. 2012). Material with low melting point yields a higher MRR, whereas a material with higher melting point and high thermal conductivity results in a lower MRR due to higher heat transfer to the nearby matrix (Lin et al. 2001). The following sections present investigation details on SEM and WSEM of aerospace alloys based on titanium, nickel and aluminium.

4.2.1.1 Titanium Alloys

Considerable research has been conducted on SEM of titanium alloys. Chen et al. (1999) used kerosene and distilled water as dielectric fluids for the SEM of Ti-6Al-4V titanium alloy and reported a higher MRR with distilled water than with kerosene, due to the formation of TiC on the surface (which has a high melting point and consumes more heat) as well as carbon deposition on the electrode. Distilled water also caused large debris formation and more micro-cracks on the surface. It was also reported that melting, vaporization and crack propagation are responsible for the material removal mechanism of titanium alloy in distilled water. Yilmaz and Okka (2010) used single-channel and multi-channel electrodes made of brass and copper for drilling holes in titanium alloy through SEM. They reported that multi-channel electrodes of both copper and brass remove less material than single-channel brass electrodes due to excessive flushing in multi-channel electrodes and low thermal conductivity of brass. MRR was found to be greatly influenced by the dielectric fluid. Gu et al. (2012) conducted comparative investigations using solid electrodes and bundled die sinking electrodes, and reported higher current endurance capacity of bundled electrodes than solid electrodes, resulting in a higher MRR. Bundled electrodes offer more effective flushing, whereas solid electrodes cause deposition of more carbon, resulting in a diminished IEG. It reduces the insulating behaviour of the dielectric fluid due to limited flushing, resulting in unstable discharge.

4.2.1.2 Nickel Alloys

Kang and Kim (2003) investigated the influence of pulse-on time on the SEM of haste alloy, and reported that MRR increases with pulse-on time up to 100 μs and then starts decreasing. This is due to an increase in discharge energy with pulse-on time; however, beyond a certain value, it renders insufficient time for the removal of debris from the IEG, causing it to deposit back on the machined surface. Ahmed and Lajis (2013) studied the SEM of Inconel 718 alloy, using a copper tool electrode, and reported that an increase in peak current increases the MRR (with the highest MRR being 34.94 mm³ per minute, when the peak current reaches 40 A, due to an increase in energy density); however, an increase in pulse-on time beyond 200 μs reduces the MRR. Mohanty et al. (2014) machined Inconel 825 alloy by SEM and reported similar findings. Kumar et al. (2010) machined Inconel 718 alloy by SEM, using graphite mixed dielectric, and observed significantly enhanced machining rate due to an increased conductivity of the dielectric. Kuppan et al. (2008) reported significant influence of peak current, duty factor and tool rotation on the MRR of Inconel 718 alloy during deep hole drilling through SEM. Bozdana et al. (2009) carried out comparative investigation on drilling of both through and blind holes in Ti-6Al-4V alloy and Inconel 718 alloy by rotary SEM using tubular hollow electrodes made of copper and brass. They reported better MRR with brass electrode than with copper electrode. Yilmaz and Okka (2010) compared the performance of SEM using single- and multi-channel brass and copper electrodes and observed a better MRR with single channel electrodes. Singh et al. (2010a) studied powder mixed SEM of haste-alloy by copper electrode and reported an influence of pulse duration, peak current and duty cycle on MRR. Bharti et al. (2010) investigated the die sinking SEM of Inconel 718 alloy, using a copper electrode, and reported a similar influence of peak current on MRR. Beri et al. (2012) studied the effects of the polarity of copper tungsten tool electrode (made of powder metallurgy) in SEM of Inconel 718, and observed that maximum MRR and minimum EWR occur with the tool electrode of positive polarity, whereas minimum surface roughness is achieved using the tool electrode with negative polarity. Rajesha et al. (2012) investigated the SEM of Inconel 718 alloy, using hollow tubular copper electrode, and reported pulse current as the most influential electrical parameter. Rajyalakshmi and Ramaiah (2013) conducted WSEM of Inconel 825 alloy, with the help of grey relational analysis, and observed increased MRR and decreased surface roughness by manipulating pulse-on time, gap voltage, servo-feed, dielectric flow rate, and wire tension.

4.2.1.3 Aluminium Alloys

One of the most important characteristics of aluminium and its alloys is their good machinability, allowing them to be machined at high speeds, thus making them suitable for the aerospace industry. Relative ease of machining them by conventional processes appears to limit the use of non-traditional machining processes for these alloys. However, there are many requirements and instances for which machining of aluminium alloys necessitates the use of SEM and SEM-based processes.

Hung et al. (1994) explored the use of SEM to machine SiC-reinforced aluminium MMC and reported that the presence of SiC reduces the MRR. Wang and Yan (2000) investigated blind hole drilling in Al_2O_3-reinforced 6061Al MMC, using Taguchi design method, and reported peak current and polarity as the most influential parameters affecting MRR, EWR and roughness. Singh et al. (2004) investigated the SEM of 10% SiC-reinforced aluminium MMC, and reported that higher MRR requires a higher current and longer pulse-on time; however, it leads to a more tapered radial overcut and surface roughness. They also reported that the flushing pressure of dielectric affects the MRR in SEM. Dhar et al. (2007) studied the influence of current, pulse-on time, and gap voltage on the SEM of aluminium-based MMC, and reported that MRR increases with current and pulse-on time, while also causing a higher radial overcut. Khan (2008) studied die-sinking SEM of aluminium alloys and mild steel, using brass and copper tool electrodes, and reported that the use of a brass tool electrode results in higher MRR of aluminium alloys, compared to a copper tool electrode. It happens due to the lower thermal conductivity of brass, which causes dissipation of less heat, thus enabling more energy to be available for a higher MRR of aluminium (which has a lower melting point than brass). Adrian et al. (2010) investigated micro-drilling in SiC- and graphite-reinforced aluminium MMC by SEM and concluded current intensity and pulse-off time to be the most influential parameters affecting MRR (which increases with an increase in their values). Satishkumar et al. (2011) studied the WSEM of SiC-reinforced Al 6063 MMC and reported that SiC reinforcement decreases the MRR of this MMC than Al6063 alloy due to the shielding provided by them against the vaporization of Al 6063 MMC. However, MRR can be increased by using longer pulse-on times, enhancing the spark intensity in turn. Velmurugan et al. (2011) machined Al 6061 MMC alloy reinforced with SiC and graphite particles by SEM and reported that MRR increases with peak current, pulse-on time, and dielectric flushing pressure, and decreases with increase in voltage. Purohit et al. (2012) used SEM to machine Al 7075 alloy reinforced with 10% SiC particles, using a brass tube electrode, and reported that MRR increases at higher speeds for electrodes of larger diameters. Singh (2012) investigated the SEM of Al_2O_3-reinforced 6061 Al alloy MMC and identified peak current as the most important parameter for optimizing SEM process parameters by grey relational analysis. Rao et al. (2014) investigated the influence of different parameters of WSEM on the MRR of 2014T6 aluminium alloy, using Taguchi method. They used a hybrid genetic algorithm for simultaneous optimization of performance measures of WSEM and reported that peak current and pulse-on time should be higher to maximize MRR.

4.2.2 Surface Integrity

Aspects of surface integrity include surface roughness, surface morphology, hardness, white recast layer, and residual stresses arising from a machining process. Surface integrity of any machined part determines its reliability. This is very important for aerospace applications. Surface integrity is significantly influenced by the machining process parameters. Surface roughness in SEM depends on the size of craters, which in turn depends on peak current, pulse duration, pulse-off time and dielectric pressure. A smaller peak current and shorter pulse duration contribute towards improved surface roughness; higher peak currents and longer pulse durations increase the discharge energy of the spark, which causes bigger and deeper melted areas, thus producing larger craters (Chen et al. 2007). The machined surface also contains spherical particles and melted drops. Increase in pulse-off time decreases the surface roughness of the SEMed surface as it provides adequate time for the surface to cool down and flush the debris and melted particles away. However, too long a pulse-off time can reduce the MRR (Rahman et al. 2011). An effective dielectric pressure, removing the molten debris, contributes towards the generation of a smooth surface through SEM. A lower dielectric pressure leads to lower heat transfer, which consequently increases the MRR and deteriorates the surface roughness. The SEMed surface experiences thermal effects and phase changes due to rapid cooling, thus making it exhibit properties different from those of the bulk material. An additional surface layer resulting from the SEM process consists of mainly white and recast layer (which is brittle in nature). This can be beneficial in enhancing wear resistance for biomedical, dental and other such applications. However, it is not desirable for aerospace applications, for it provides sites for generation and propagation of micro- and macro-cracks, and may help in spreading cracks towards the bulk material. Another important subsurface property of SEMed materials is residual stress arising due to non-homogeneous temperature distribution and rapid cooling effect in SEM and WSEM processes. Reduction in fatigue strength due to residual stresses (particularly tensile) can be overcome by appropriate heat treatment. Micro-hardness of the SEMed surface increases up to a distance of 100 μm from the top and remains constant at further depths. This is due to the formation of oxides and carbides on the SEMed surface, caused by the dissociation of the dielectric at the elevated temperature and due to the transfer of the electrode material.

Speed and tension of wire are critical parameters in WSEM. Low wire speeds cause more melting and removal of material, causing higher surface roughness. Similarly, an extremely low or too high wire tension can cause vibrations or breakage of the wire. The following paragraphs summarize the research on surface integrity aspects of aerospace alloys in SEM and WSEM processes.

4.2.2.1 Titanium Alloys

Chen et al. (1999) investigated the effect of dielectric type on the SEM of Ti-6Al-4V alloy and found that SEM with distilled water causes more cracks than kerosene. This is due to the higher thermal conductivity and higher cooling rate provided by distilled water. They also investigated the effect of pulse-on time on recast layer thickness and observed that it increases with pulse-on time in the range of 3-6 μs; however, when pulse duration exceeds 12 μs, the recast layer thickness starts decreasing due to the higher impact force of the dielectric fluid, which flushes away the machined products from the IEG. Lin et al. (2000) also confirmed these observations. Hasçalık and Çaydaş (2007) assessed the influence of peak current and pulse duration on surface roughness and recast layer thickness in the SEM of Ti-6Al-4V alloy, and found that both increase with peak current, due to the generation of deeper and larger craters by melting and solidification of more material. They observed minimum thickness of recast layer occurring at a peak current of 3A. No structural changes, except the solidified layer, were observed for increased values of peak current. Ekmekci (2007) studied residual stresses and recast layer in SEM and observed that exposure to elevated temperatures and increased quenching effect cause metallurgical alteration. Fonda et al. (2008) reported that surface morphology generated during SEM is influenced by material properties and machining conditions and that it varies with duty cycle. They found that duty cycle values in the range of 3-7% cause individual craters to be formed on the surface (Fig. 4.3a), whereas values in the range of 20% to 50% cause individual craters to blend together (Fig. 4.3b) due to higher temperatures and insufficient time available for the SEMed surface to cool down before the next cycle of sparks starts.

Kibria et al. (2010) investigated the effect of dielectric on recast layer formation during the drilling of micro-holes in Ti-6Al-4V alloy by micro-SEM process and found that the use of distilled water as a dielectric causes the formation of a recast layer thinner than one obtained with kerosene. This is due to the higher cooling rate provided by distilled water, which causes a rapid transfer of the generated heat to the molten material and the dielectric, thus reducing the likelihood of debris adherence. For both the dielectric fluids, they also observed an increase in the recast layer thickness with an increase in pulse-on time. This was attributed to the larger machining time, enabled by a longer pulse-on time, resulting in more debris which is likely to adhere to and re-solidify on the SEMed surface upon cooling. Adding an additive to the dielectric can flush the debris more efficiently, resulting in a thinner recast layer than with pure distilled water. Stráský et al. (2011) investigated the SEM effects of microstructure on the fatigue behaviour of

|IMS   2007/06/07 11:08   x200   500 um|IMS   2007/06/07 11:36   x200   500 um|
(a)   (b)

Fig. 4.3: SEM images of Ti-6Al-4V alloy using (a) 7% duty factor and (b) 50% duty factor (Fonda et al. 2008, © Elsevier. Reprinted with permission)

Ti-6Al-4V alloy for orthopaedic applications. Ndaliman et al. (2013) studied the micro-hardness of Ti-6Al-4V alloy in SEM process and reported that the combination of Cu-TaC electrode and urea in the dielectric has an alloying tendency with the machined surface, which enhances the micro-hardness of the surface. Peak current also influences micro-hardness. Thesiya et al. (2014) studied recast layers and heat affected zones (HAZ) in Ti-4Al-6V alloy in SEM process, using copper and graphite tool electrodes with positive polarity. They found that the tool electrode material and its polarity are the most influential factors for the recast layer, and reported that the thickness of the recast layer and the HAZ increases with voltage and current for a copper tool electrode, whereas graphite shows a opposite trend.

Lin et al. (2001) reported that the hardness of TiNi and TiNiCu shape memory materials increased from 200Hv to 750Hv during SEM due to the formation of TiO_2 and $TiNiO_3$ (Fig. 4.4). Chen et al. (2007) also conducted similar investigations and reported an increase in hardness, up to 100 μm depth from the top of the SEMed surface, due to the formation of oxides such as Cr_2O_3, ZrO_2, TiO_2, $TiNiO_3$ and TiC in the recast layer of $Ti_{50}Ni_{49.5}Cr_{0.5}$ and $Ti_{35.5}Ni_{49.5}Zr_{15}$ shape memory alloys. They concluded that this hardness increment is more pronounced with higher values of pulse duration and peak current. Aspinwall et al. (2008) found insignificant influence of tool electrodes on the hardness of the WSEMed surface though repeated machining pass generates better hardness characteristics. Hsieh et al. (2009) investigated change in hardness with distance from top of the WSEMed surfaces of $Ti_{35.5}Ni_{49.5}Zr_{15}$ and $Ti_{50}Ni_{49.5}Cr_{0.5}$ shape memory alloys. They found that hardness on the top surface can be as high as 875Hv for $Ti_{35.5}Ni_{49.5}Zr_{15}$ alloy and 807Hv for $Ti_{50}Ni_{49.5}Cr_{0.5}$ alloy.

Kuriakose and Shunmugam (2004) studied the surface characteristics of Ti-4Al-6V alloy in WSEM, using zinc-coated brass and pure brass wires, and observed that the coated wire gives a more uniform surface than the

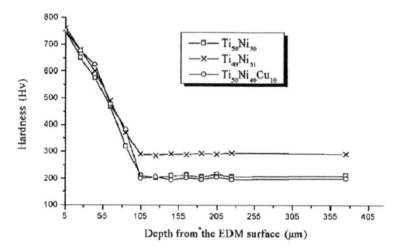

Fig. 4.4: Change in the hardness of different shape memory materials, at various distances from the WSEMed surface (peak current = 25A, pulse duration = 12 µs) (Lin et al. 2001, © Springer. Reprinted with permission)

uncoated wire due to the absence of variable thermal load. Study by Alias et al. (2012) on the WSEM of Ti-4Al-6V alloy suggested the use of an optimum value of wire speed for higher MRR and lower surface roughness and kerf width.

4.2.2.2 Nickel Alloys

Extensive investigations have been conducted on SEM-induced surface integrity of different nickel-based alloys. Kang and Kim (2003) investigated the surface characteristics of heat resistant nickel alloy and found that a good surface is produced with shorter pulse durations due to the effective removal of debris from the IEG and that larger pulse durations increase the number of micro-cracks and thickness of the altered material zone. They also reported that the generated micro-cracks do not vanish even after heat treatment and that the recast layer offers reduced hardness due to annealing caused by the release of residual stresses after heat treatment. Refined grains appear due to recrystallization during heat treatment. Klocke et al. (2004) studied the effect of mixing powder in the dielectric for the SEM of Inconel 718, and observed changes in the composition and morphology of the recast layer produced. Kuppan et al. (2008) investigated deep-hole drilling in Inconel 718 by SEM process and concluded pulse duration and peak current to be relatively more important parameters affecting surface roughness. Newton et al. (2009) studied the effect of WSEM process parameters on the formation and the characteristics of recast layer in Inconel 718, and observed increased thickness of the recast layer with increased discharge energy, discharge current, and pulse duration. This recast layer experiences not only lower hardness, but also lower elastic modulus, compared to bulk material. Rajesha et al. (2010) investigated the surface integrity of Inconel 718 alloy by SEM process and reported poor surface integrity conditions due to higher values of pulse duration and peak current, and smaller pulse-off time. They also concluded that duty factor and composition of the recast layer have substantial influence on crack propagation. Yilmaz and Okka (2010) conducted a comparative study of SEM of Inconel 718 and Ti-6Al-4V alloy using single- and multi-channel electrodes made of brass and copper. They reported better surface roughness for multi-channel electrodes. Kumar et al. (2010) investigated the SEM of Inconel 718 alloy, using graphite-mixed dielectric, and copper electrode formed using powder metallurgy. They observed a decreasing trend of surface roughness up to 14A of peak current; however, this trend was reversed with further increase in the peak current, due to the material deposition. Prabhu and Vinayagam (2011) did an atomic force microscopy (AFM) analysis of SEMed nickel alloy Inconel 825 by mixing multi-wall carbon nano-tubes and single-wall carbon nano-tubes (Prabhu and Vinayagam, 2013) in the dielectric. They reported better surface finish at low currents, formation of nano-cracks, and a direct relationship between the depth of cracks and surface roughness with input power. Ahmad and Lajis (2013) studied the effects of pulse-on-time and peak current on the SEM of Inconel 718, using a copper electrode, and found that surface roughness increases with increase in these parameters. This was due to the generation of larger craters. They reported surface roughness as 8.53 μm at 20 A peak current and 200 μs pulse duration. Mohanty et al. (2014) obtained similar results for Inconel 825. Perveen and Jahan (2016) conducted investigations on micro-SEM of nickel-based Haste alloy and reported an increase in the crater size with increased discharge energy.

4.2.2.3 Aluminium Alloys

Hung et al. (1994) studied the SEM of SiC-reinforced cast Al alloy and reported that current has a dominant effect on surface roughness, and that the unmolten SiC, which drops out without vaporizing, resolidifies on the surface to form the recast layer. They also observed that the HAZ and the recast layer were crack-free. Yan et al. (2000) investigated rotary SEM of Al 6061 MMC reinforced with alumina particles, and reported that a rougher surface is generated if the composite contains 20% alumina particles or at higher speed of tool rotation. Yan et al. (2005a) studied the SEM of the same MMC and found that surface roughness depends on the volume fraction of reinforcement particles, whereas servo voltage, pulse-on time and pulse-off time have little impact on the same. Patel et al. (2009) studied the influence of SEM process parameters on MRR and damage to ceramic matrix composite (CMC) consisting of Al_2O_3 as the matrix material, reinforced with SiC and TiC. Satishkumar et al. (2011) investigated the SEM of SiC-reinforced Al6063 MMC and Al6063 alloy using Taguchi's L_9 orthogonal array and reported that the reinforcement of Al6063 by SiC increases its surface roughness compared to Al6063 alloy. Selvakumar et al. (2014) used Taguchi's L_9 orthogonal array to study the effects of pulse-on time, pulse-off time, discharge current and wire tension on the performance of WSEM of 5083 aluminium alloy. They used the concept of pareto-optimality to determine the optimum values of the considered WSEM process parameters and reported that surface roughness is not influenced by pulse-off time and wire tension, and that cutting speed is independent of wire tension. Rao et al. (2014) investigated the WSEM of aluminium 2014T6 alloy to optimize the WSEM parameters for MRR and surface roughness. Their results revealed that shorter pulse durations, smaller peak currents, and higher pulse-off times result in an optimum surface roughness. Wire tension influences surface roughness, due to its effect on wire vibrations; a higher wire tension results in a reduction in surface roughness. Measurement of recast layer thickness reveals that it varies from 5 μm to 50 μm, based on the choice of WSEM parameters. Pramanik et al. (2015) studied the WSEM of 6061 aluminium alloy and reported tapered WSEMed slits at all machining conditions, due to erosion of the wire electrode at higher heat inputs. One of the main problems associated with WSEM of aluminium alloys is the edge start due to wire breakage caused by the formation of an oxide layer on the surface of the aluminium alloy. This requires a cleaning process before the start of WSEM. Increasing pulse-off time is also helpful in avoiding wire breakage.

4.2.3 Tool Wear Ratio

Tool wear ratio (TWR) is defined as the ratio of material removed from the tool electrode to that removed from the workpiece (that is, MRR). Notwithstanding the fact that electrode wear naturally occurs in SE-based processes, its value should be kept as low as possible, for it directly affects surface quality and productivity of such processes. Relevant research and their findings in this context have been detailed in subsequent sub-sections.

4.2.3.1 Titanium Alloys

Chen et al. (1999) studied the SEM of Ti-6Al-4V alloy and found that electrode wear increases with pulse-on time. They also found that the use of distilled water as a dielectric results in lower TWR than kerosene, due to the formation of oxides of lower melting point than that obtained with kerosene. Therefore, the use of distilled water as a dielectric results in a higher MRR and better precision than kerosene. Lin et al. (2000) carried out a similar investigation using SEM combined with ultrasonic machining (USM) for Ti-6Al-4V alloy and reported findings similar to those of Chen et al. (1999). Chen et al. (2007) investigated the TWR of $Ti_{50}Ni_{49.5}Cr_{0.5}$ and $Ti_{35.5}Ni_{49.5}Zr_{15}$ shape memory alloys in the SEM process and observed that TWR increases with pulse duration up to a certain value and starts decreasing thereafter. This is due to the fact that longer pulse durations increase the discharge energy of a spark, causing

faster melting and vaporization of materials from the tool electrode and
the workpiece, but pulse duration beyond a certain value starts expanding
the plasma channel, rendering insufficient energy density, thus causing the
accumulation of some molten debris in the IEG, which in turn reduces the
TWR. Yilmaz and Okka (2010) studied hole drilling in Ti alloy by SEM, using
single- and multi-channel electrodes made of copper and brass, and found that
a multi-channel brass electrode experiences more wear than other electrodes
and that single- and multi-channel copper tool electrodes experience similar
wear as shown in Fig. 4.5. This phenomenon can be explained with the
help of volumetric ratio of electrodes – 1.21 and 1.06 for brass and copper
electrodes respectively. Brass multi-channel electrodes have lower volume
than single-channel electrodes. Therefore, it wears about 1.2 times faster.
Kibria et al. (2010) investigated micro-hole drilling in Ti-6Al-4V by μ-SEM,
using deionized water and kerosene as the dielectrics and obtained a higher
TWR with deionized water than with kerosene. This was attributed to the
decomposition of kerosene into carbon which got deposited on the surface
of the tool electrode, forming a protective layer and thus reducing tool wear.
They also observed that kerosene mixed with B_4C particles results in less tool
wear than pure kerosene, for peak currents of 0.5A and 1A, due to increase
in the number of carbon particles being released by both B_4C powder and
kerosene. Moreover, tool wear is lower for higher current (1A) due to the
decomposition of more kerosene at higher discharge energies. Furthermore,
TWR obtained with deionized water mixed with B_4C powder is lower for a
peak current of 2A than for a peak current of 1.5A. Thus, it can be inferred
that addition of powdered material can help in reducing TWR, due to the
deposition of carbon particles.

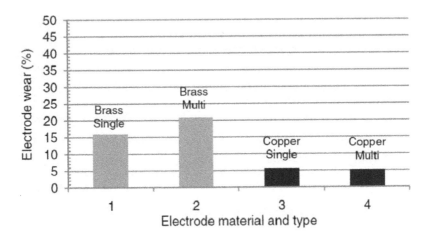

Fig. 4.5: Wear of single- and multi-channel tool electrodes made of brass and copper,
during SEM of Ti-6Al-4V (Yilmaz and Okka 2010 © Springer Nature. Reprinted with
permission)

4.2.3.2 Nickel Alloys

Kang and Kim (2003) studied the effect of SEM parameters on the TWR of nickel alloy and observed a gradual increase in the TWR with pulse-on time, due to the adherence of a wear-resistant layer of carbon formed due to the decomposition of the dielectric fluid and the workpiece material. Their EDS analysis evidently supported these findings. Arun et al. (2009) studied magnetic force assisted SEM of Inconel 800 and reported enhancements in MRR, TWR and surface characteristics, compared to conventional SEM. Bharti et al. (2010) also studied the effect of pulse duration on the TWR of Inconel 718. Yilmaz and Okka (2010) reported that a single-channel electrode yields a lower TWR than a multi-channel electrode for drilling holes in a nickel-based alloy through SEM. Singh et al. (2010a) studied the influence of peak current and gap voltage on the TWR of haste alloy using aluminium powder mixed SEM, and reported that a higher value of peak current and lower value of gap voltage results in a lower TWR.

Ahmed and Lajis (2013) studied effect of pulse-on time and peak current and observed increase in TWR with peak current due to increase in discharge energy and decrease with increase of pulse duration due to carbon deposition. Their observations are in good agreement with Chen et al. (2007).

4.2.3.3 Aluminium Alloys

Singh et al. (2004) reported larger tool wear for cast aluminium MMC at higher values of current. Dhar et al. (2007) investigated the TWR of 10wt% SiC$_P$-reinforced Al-4Cu-6Si alloy MMC and found that TWR increases with increase in current, due to the higher thermal loading of both the electrodes, which in turn causes more material removal. They also observed that an increase in pulse duration and gap voltage reduces the TWR initially; however, beyond their optimum values, TWR starts increasing with both the parameters. Khan (2008) studied MRR and TWR for aluminium and mild steel in die-sinking SEM using electrodes made of copper and brass and revealed that TWR of aluminium is lower than steel. He also observed that copper electrodes experience less TWR than brass electrodes, which is mainly attributed to the high thermal conductivity of copper. Adrian et al. (2010) studied TWR of SiC-reinforced MMC of aluminium in micro-drilling by SEM, and reported discharge current and pulse-on time as the most influential parameters, and suggested the use of low pulse-on times for lower tool wear. Velmurugan et al. (2011) investigated the MMC of Al6061 reinforced with graphite and SiC and found increased tool wear due to increase in current and voltage, and decrease in tool wear with increase in pulse-on time and flushing pressure.

4.3 Tool Wear Compensation Techniques

SEM process on both micro- and macro-scale suffers with problems associated with wear of tool electrode and low material removal efficiency. If wear of the tool electrode becomes severe, then it affects the depth and the shape of the SEMed products significantly. Therefore, it is essential to compensate the tool wear during the SEM process. There are two types of wear, namely, non-uniform (or conventional) wear and uniform wear. Non-uniform wear or conventional wear involves wear of the tool electrode both at its bottom and its corners (Fig. 4.6a). Bottom wear changes its surface depth, whereas corner wear imparts geometrical inaccuracy to it. This type of wear is very difficult to compensate. In uniform wear method (UWM), the tool retains its original shape during SEM because tool wear occurs only at its bottom (Fig. 4.6b). Uniform tool wear can be insured if the material is removed layer by layer, with assistance from a simple electrode whose tool path is generated based on the uniform tool wear where layer depth is less than machining gap which confines the discharges only to the bottom of the tool. Figure 4.6c shows the tool path for manufacturing a square shape.

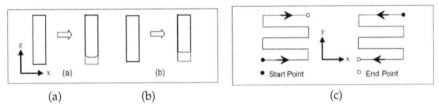

(a)   (b)   (c)

Fig. 4.6: Wear of tool electrode in SEM process: (a) conventional non-uniform wear, (b) uniform wear, and (c) tool path for square shape (Narasimhan et al. 2005, © Elsevier. Reprinted with permission)

Many techniques such as wear forecasting, apprehending, modelling, and tool wear compensation have been proposed to avoid or minimize the effects of electrode tool wear on the geometry and accuracy of the SEMed product (Narasimhan et al. 2005). Uniform wear of the tool electrode can be compensated by using a compensation equation in which worn length is added to the tool length after the removal of each layer (Yu et al. 1998). Additionally, UWM has the capability to be integrated with CAD/CAM in order to facilitate SEM of 3D features (Rajurkar and Yu 2000). In linear compensation method (LCM), wear-caused reduction in tool length is continuously compensated by moving the tool in small increments along its feed direction. But this technique needs a lot of experimentation to find the ratio of tool feed to the distance travelled by the tool. Moreover, it is not suitable for manufacturing complicated 3D parts (Yuzawa et al. 1997). Therefore, changes in tool size and shape can be monitored using a digital imaging system, and compensation can be done accordingly. Bleys et al. (2012) proposed a tool wear monitoring technique in which occurrence of sparks and rate of spark generation are monitored and documented by the system and tool compensation is done accordingly. However, there are certain difficulties associated with the analysis of recorded data and calculation of appropriate compensation, which limits its application. Yu et al. (2010) proposed to combine both LCM and UWM to enhance MRR as well as to reduce TWR and surface roughness.

4.4 Modelling of Spark Erosion of Aerospace Alloys

SEM is a very complex machining process as it involves interaction among several aspects of science and engineering, such as electrical, thermal, mechanical, plasma, fluid dynamics, etc. Due to its stochastic nature and the interaction of these aspects, its mechanism of material removal has not been fully understood yet. Therefore, development of models of SEM, which are capable of predicting process outputs, is of paramount importance. These models not only make it possible to understand the physics behind the SEM process, but also help in selecting optimum or sub-optimum SEM parameters to get the required surface integrity and tolerance with an increased MRR. The process of material removal and surface generation in SEM is the superposition of individual sparks, which are different from each other, due to the transient nature of the SEM process. This transient nature is due to temporal and spatial variations in gap distance, plasma pressure, temperature and chemical composition of the dielectric fluid. However, understanding the interaction of single sparks with the workpiece material is crucial to understand the physics behind this process. A generalized sequence of activities involved in the development of a model of the SEM process is presented in Fig. 4.7.

Since the nature of material removal in SEM is thermal due to melting and vaporization, thermal modelling is apparently more significant (Hinduja and Kunieda 2013). It has been shown by a number of researchers as well that a thermal model of SEM process can be used to predict MRR. An electro-thermal model developed by Snoeys and Van Dijck (1971) considered the heat source as a semi-infinite cylinder and assumed equal energy distribution between both the electrodes. Beck (1981a, 1981b) assumed constant thermal properties as well as constant heat flux to study the heat distribution at the cathode. However, they did not take into account convective heat transfer and energy transferred to the cathode. Three-dimensional transient heat conduction model of SEM was developed by Panda and Bhoi (2001), with the assumption of an expanded plasma channel. The limitation of this model lies in assuming constant radius of spark under all discharge conditions. It is worth mentioning that the aforementioned researches considered incorporating the effects of neither plasma flushing efficiency (PFE) nor phase change in their models. Yeo et al. (2008) carried out a critical comparison of various thermal models of SEM in predicting temperature distribution, crater size, and MRR. Their comparative assessment of the ratio of numerical to experimental data for MRR (for discharge energies in the range of 0.33 to 952 mJ) showed that DiBitonto's model (DiBitonto et al. 1989) matches well with experimental data, yielding the closest proximity of 1.2-46.1 MRR when compared with other models.

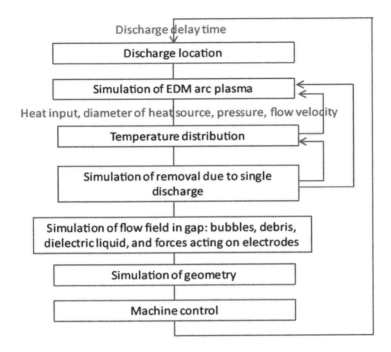

Fig. 4.7: Generalized sequence of activities to be considered for developing a model of the SEM process (Hinduja and Kunieda, 2013 © Elsevier. Reprinted with permission)

Finite element analysis (FEA) has been extensively used to model the SEM process. A finite element model of EDM process was developed by Yadav et al. (2002) to estimate the temperature field and thermal stresses, considering the heat source in the SEM process having Gaussian distribution. They calculated temperature distribution using a distribution factor of 0.42. Their model has some limitations, namely, non-consideration of phase change, plasma flushing efficiency, temperature-dependence of material properties, and consideration of constant spark radius. Kalajahi et al. (2013) developed a better FEA-based model of the SEM process by including factors such as plasma flushing efficiency (PFE), latent heat of melting, Gaussian heat flux, temperature dependent material properties, and energy distribution. Kumar et al. (2014) developed a numerical model for single discharge WSEM of Inconel 718, and subsequently generalized it to the multi-spark case by calculating the number of pulses. Using the generalized model, they obtained MRR.

Other approaches which have been widely used to model and/or optimize the parameters of SEM and WSEM processes include data mining, artificial neural network, response surface methodology (RSM), regression analysis, grey relational analysis, and Taguchi's orthogonal array employing signal-to-noise (S/N) ratios. Analysis of variance (ANOVA) and S/N ratios have been used to determine significant input parameters of SEM and WSEM processes. Chiang and Chang (2006) used grey relational analysis to optimize WSEM parameters during the machining of 6061 alloy, considering performance measures of MRR and maximum surface roughness value. Their study indicated that grey relational analysis is a suitable tool for optimization of WSEM process parameters. Sarkar et al. (2005) carried out an L_{18} Taguchi experimental design to investigate single pass cutting in the WSEM of γ-titanium aluminide alloy. They constructed an additive model to evaluate the influence of control factors. The process was optimized using constrained and pareto-optimization algorithms. The achieved pareto-optimal solution set (with twenty solutions) was very effective and capable of acting as a guideline for WSEM of γ-titanium aluminide alloys.

Data mining approach based on machine learning was used by Kuriakose et al. (2003) to model WSEM of Ti-6Al-4V titanium alloy; they studied the influence of various WSEM parameters on cutting speed and surface finish. Aggarwal et al. (2015) used RSM for parametric modelling and optimization of WSEM of Inconel 718, considering pulse duration, pulse-off time, discharge current, discharge voltage, wire feed rate and wire tension as input variables, and MRR and surface roughness as performance indicators. Their results revealed that the factors affecting MRR, in decreasing order of their influence, are pulse-duration, discharge voltage and pulse-off time. Pulse duration and discharge voltage were found to be the major factors affecting the surface roughness of the WSEMed product.

4.5 Summary

Aerospace materials are meant to operate in extremely high pressure and temperature conditions, while maintaining a high specific weight to reduce the consumption of fuel as well as the operational cost. It is evident from the literature review presented in this chapter that profound investigations have been conducted in the past 2-3 decades to improve material processing, machining and manufacturing techniques of aerospace materials. Machinability of aerospace materials (even those considered difficult-to-machine) has been improved with the use of SEM and WSEM. However, SEM suffers from limitations such as tool electrode wear, long machining times, and the presence of heat affected zones. Therefore, identifying optimum parameters and evolving hybrid machining to overcome the limitations of a single machining process offer significant potential for improved machinability of aerospace materials. Continuous efforts towards discovering and developing newer aerospace alloys need continuous improvement of SEM to achieve the best possible (optimized) performance parameters. Future research in this direction should focus on the use of assistive SEM processes for newer aerospace alloys. Application of nano-SEM can be another interesting direction for future research. Since safety factors for the aeroplane industry cannot be traded-off in any case, with the advent of economical airlines, extensive research is needed on the feasibility of economical processing of aerospace materials to manufacture economical aeroplanes.

References

Adrian, I., Eugen, A. and Florin, N. 2010. A study about micro-drilling by electrical discharge method of an Al/SiC hybrid composite. International Journal of Academic Research, 2(3). 6-13.

Aggarwal, V., Khangura, S.S. and Garg, R.K. 2015. Parametric modeling and optimization for wire electrical discharge machining of Inconel 718 using response surface methodology. The International Journal of Advanced Manufacturing Technology, 79(1-4): 31-47.

Ahmad, S. and Lajis, M.A. 2013. Electrical discharge machining (EDM) of Inconel 718 by using copper electrode at higher peak current and pulse duration. Conference Proceeding of IOP Conference Series: Materials Science and Engineering.

Alias, A., Abdullah, B. and Abbas, N.M. 2012. Influence of machine feed rate in WEDM of titanium Ti-6Al-4V with constant current (6A) using brass wire. Procedia Engineering, 41: 1806-1811.

Arun, M.B., Karthik, K., Soundararajan, R. and Palanisamy, A. 2009. Characteristics of magnetic force-assisted electric discharge machining on Inconel 800. Proceedings of 6th International Conference on Precision, Meso, Micro and Nano Engineering (COPEN 2009), pp. G7–G11.

Aspinwall, D., Soo, S., Berrisford, A. and Walder, G. 2008. Workpiece surface roughness and integrity after WEDM of Ti-6Al-4V and Inconel 718 using minimum damage generator technology. CIRP Annal – Manufacturing Technology, 57(1): 187-190.

Beck, J.V. 1981a. Large time solutions for temperatures in a semi-infinite body with a disk heat source. International Journal of Heat and Mass Transfer, 24(1): 155-164.

Beck, J.V. 1981b. Transient temperatures in a semi-infinite cylinder heated by a disk heat source. International Journal of Heat and Mass Transfer, 24(10): 1631-1640.

Beri, N., Pungotra, H. and Kumar, A. 2012. To study the effect of polarity and current during electric discharge machining of Inconel 718 with CuW powder metallurgy electrode. Proceedings of the National Conference on Trends and Advances in Mechanical Engineering, YMCA University of Science and Technology, Faridabad, Haryana.

Bharti, P.S., Maheshwari, S. and Sharma, C. 2010. Experimental investigation of Inconel 718 during die-sinking electric discharge machining. International Journal of Engineering Science and Technology, 2(11): 6464-6473.

Bleys, P., Kruth, J.P., Lauwers, B., Zryd, A., Delpretti, R. and Tricarico, C. 2002. Real-time tool wear compensation in milling EDM. CIRP Annal – Manufacturing Technology, 51(1): 157-160.

Boothroyd, G. and Knight, W. 1989. Fundamentals of Machining and Machine Tools. Marcel-Dekker. New York.

Bozdana, A., Yilmaz, O., Okka, M. and Filiz, I. 2009. A comparative experimental study on fast hole EDM of Inconel 718 and Ti-6Al-4V. Proceeding of 5th International

Conference and Exhibition on Design and Production of Machines and Dies/molds.

Chen, S., Yan, B. and Huang, F. 1999. Influence of kerosene and distilled water as dielectrics on the electric discharge machining characteristics of Ti-6A1-4V. Journal of Materials Processing Technology, 87(1): 107-111.

Chen, S., Hsieh, S., Lin, H., Lin, M. and Huang, J. 2007. Electrical discharge machining of TiNiCr and TiNiZr ternary shape memory alloys. Materials Science and Engineering: A, 445: 486-492.

Chiang, K.T. and Chang, F.P. 2006. Optimization of the WEDM process of particle-reinforced material with multiple performance characteristics using grey relational analysis. Journal of Materials Processing Technology, 180(1): 96-101.

Dhar, S., Purohit, R., Saini, N., Sharma, A. and Kumar, G.H. 2007. Mathematical modeling of electric discharge machining of cast Al-4Cu-6Si alloy–10wt.% SiC$_P$ composites. Journal of Materials Processing Technology, 194(1): 24-29.

DiBitonto, D.D., Eubank, P.T., Patel, M.R. and Barrufet, M.A. 1989. Theoretical models of the electrical discharge machining process. I: A simple cathode erosion model. Journal of Applied Physics, 66(9): 4095-4103.

Ekmekci, B. 2007. Residual stresses and white layer in electric discharge machining (EDM). Applied Surface Science, 253(23): 9234-9240.

Erden, A. 1983. Effect of materials on the mechanism of electric discharge machining. ASME Transactions: Journal of Engineering Materials and Technology, 105: 132-138.

Ezugwu, E.O., Wang, Z.M. and Machado, A.R. 1999. The machinability of nickel-based alloys: A review. Journal of Materials Processing Technology, 86(1): 1-16.

Fonda, P., Wang, Z., Yamazaki, K. and Akutsu, Y. 2008. A fundamental study on Ti-6Al-4V thermal and electrical properties and their relation to EDM productivity. Journal of Materials Processing Technology, 202(1): 583-589.

Gadalla, A. and Tsai, W. 1989. Machining of WC-Co composites. Materials and Manufacturing Processes, 4(3): 411-423.

Gangadhar, A., Shunmugam, M. and Philip, P. 1992. Pulse train studies in EDM with controlled pulse relaxation. International Journal of Machine Tools and Manufacture, 32(5): 651-657.

Gu, L., Li, L., Zhao, W. and Rajurkar, K.P. 2012. Electrical discharge machining of Ti6Al4V with a bundled electrode. International Journal of Machine Tools and Manufacture, 53(1): 100-106.

Hasçalık, A. and Çaydaş, U. 2007. A comparative study of surface integrity of Ti-6Al-4V alloy machined by EDM and AECG. Journal of Materials Processing Technology, 190(1): 173-180.

Heinz, A., Haszler, A., Keidel, C., Moldenhauer, S., Benedictus, R. and Miller, W.S. 2000. Recent development in aluminium alloys for aerospace applications. Materials Science and Engineering: A, 280(1): 102-107.

Hinduja, S. and Kunieda, M. 2013. Modelling of ECM and EDM processes. CIRP Annal – Manufacturing Technology, 62(2): 775-797.

Hsieh, S., Chen, S., Lin, H., Lin, M. and Chiou, S. 2009. The machining characteristics and shape recovery ability of Ti–Ni–X (X = Zr, Cr) ternary shape memory alloys using the wire electro-discharge machining. International Journal of Machine Tools and Manufacture, 49(6): 509-514.

Hung, N., Yang, L. and Leong, K. 1994. Electrical discharge machining of cast metal matrix composites. Journal of Materials Processing Technology, 44(3-4): 229-236.

Jameson, E. C. (1983). Electrical discharge machining: tooling, methods, and applications. Society of Manufacturing Engineers.

Jameson, E.C. 2001. Electrical Discharge Machining, Marketing Service Division. Society of Manufacturing Engineers, Dearborn, Michigan, USA.

Kalajahi, M.H., Ahmadi, S.R. and Oliaei, S.N.B. 2013. Experimental and finite element analysis of EDM process and investigation of material removal rate by response surface methodology. The International Journal of Advanced Manufacturing Technology, 69(1-4): 687-704.

Kalpajian, S. and Schmid, S. 2003. Material Removal Processes: Abrasive, Chemical, Electrical and High-energy Beam. Manufacturing Processes for Engineering Materials. Prentice Hall, New Jersey.

Kandpal, B.C., Kumar, J. and Singh, H. 2015. Machining of aluminium metal matrix composites with electrical discharge machining – A review. Materials Today: Proceedings, 2(4): 1665-1671.

Kang, S.H. and Kim, D.E. 2003. Investigation of EDM characteristics of nickel-based heat resistant alloy. KSME International Journal, 17(10): 1475-1484.

Khan, A.A. 2008. Electrode wear and material removal rate during EDM of aluminum and mild steel using copper and brass electrodes. The International Journal of Advanced Manufacturing Technology, 39(5): 482-487.

Kibria, G., Sarkar, B., Pradhan, B. and Bhattacharyya, B. 2010. Comparative study of different dielectrics for micro-EDM performance during microhole machining of Ti-6Al-4V alloy. The International Journal of Advanced Manufacturing Technology, 48(5): 557-570.

Klocke, F., Lung, D., Antonoglou, G. and Thomaidis, D. 2004. The effects of powder suspended dielectrics on the thermal influenced zone by electrodischarge machining with small discharge energies. Journal of Materials Processing Technology, 149(1): 191-197.

Krar, S.F. and Check, A.F. 1997. Electrical Discharge Machining: Technology of Machine Tools. Glencoe/McGraw-Hill, New York.

Kumar, A., Bagal, D.K. and Maity, K. 2014. Numerical modeling of wire electrical discharge machining of superalloy Inconel 718. Procedia Engineering, 97: 1512-1523.

Kumar, A., Maheshwari, S., Sharma, C. and Beri, N. 2010. Realizing potential of graphite powder in enhancing machining rate in AEDM of nickel based super alloy 718. Proceedings of the International Conference on Advanced in Mechanical Engineering.

Kunieda, M., Lauwers, B., Rajurkar, K. and Schumacher, B. 2005. Advancing EDM through fundamental insight into the process. CIRP Annal – Manufacturing Technology, 54(2): 64-87.

Kuppan, P., Rajadurai, A. and Narayanan, S. 2008. Influence of EDM process parameters in deep hole drilling of Inconel 718. The International Journal of Advanced Manufacturing Technology, 38(1-2): 74-84.

Kuriakose, S., Mohan, K. and Shunmugam, M.S. 2003. Data mining applied to wire-EDM process. Journal of Materials Processing Technology, 142(1): 182-189.

Kuriakose, S. and Shunmugam, M. 2004. Characteristics of wire-electro discharge machined Ti6Al4V surface. Materials Letters, 58(17): 2231-2237.

Lee, T. and Lau, W. 1991. Some characteristics of electrical discharge machining of conductive ceramics. Materials and Manufacturing Processes, 6(4): 635-648.

Liao, Y.S., Huang, J.T. and Su, H.C. 1997. A study on the machining parameters optimization of wire electrical discharge machining. Journal of Materials Processing Technology, 71(3): 487-493.

Lin, Y.C., Yan, B.H. and Chang, Y.S. 2000. Machining characteristics of titanium alloy (Ti-6Al-4V) using a combination process of EDM with USM. Journal of Materials Processing Technology, 104(3): 171-177.

Lin, H., Lin, K. and Cheng, I. 2001. The electro-discharge machining characteristics of TiNi shape memory alloys. Journal of Materials Science, 36(2): 399-404.

Manjaiah, M., Narendranath, S. and Basavarajappa, S. 2014. A review on machining of titanium based alloys using EDM and WEDM. Reviews on Advanced Materials. Science, 36(2): 89-111.

McGeough, J.A. 1988. Advanced Methods of Machining. Springer Netherlands.

Mohanty, A., Talla, G. and Gangopadhyay, S. 2014. Experimental investigation and analysis of EDM characteristics of Inconel 825. Materials and Manufacturing Processes, 29(5): 540-549.

Narasimhan, J., Yu, Z. and Rajurkar, K.P. 2005. Tool wear compensation and path generation in micro and macro EDM. Journal of Manufacturing Processes, 7(1): 75-82.

Ndaliman, M.B., Khan, A.A. and Ali, M.Y. 2013. Influence of electrical discharge machining process parameters on surface micro-hardness of titanium alloy. Proceedings of the Institution of Mechanical Engineers, Part B: Journal of Engineering Manufacture, 227(3): 460-464.

Newton, T.R., Melkote, S.N., Watkins, T.R., Trejo, R.M. and Reister, L. 2009. Investigation of the effect of process parameters on the formation and characteristics of recast layer in wire-EDM of Inconel 718. Materials Science and Engineering: A, 513: 208-215.

Panda, D.K. and Bhoi, R.K. 2001. Developing transient three dimensional thermal models for electro discharge machining of semi infinite and infinite solid. Journal of Materials Processing and Manufacturing Science, 10: 71-89.

Patel, K., Pandey, P.M. and Rao, P.V. 2009. Surface integrity and material removal mechanisms associated with the EDM of Al_2O_3 ceramic composite. International Journal of Refractory Metals and Hard Materials, 27(5): 892-899.

Perveen, A. and Jahan, M. 2016. An experimental study on the effect of operating parameters during the micro-electro-discharge machining of Ni based alloy. World Academy of Science, Engineering and Technology. International Journal of Chemical, Molecular, Nuclear, Materials and Metallurgical Engineering, 10(11): 1367-1373.

Peters, M., Kumpfert, J., Ward, C.H. and Leyens, C. 2003. Titanium alloys for aerospace applications. Advanced Engineering Materials, 5(6): 419-427.

Peters, M. and Leyens, C. 2009. Aerospace and space materials. Materials Science and Engineering, 3: 1-11.

Prabhu, S. and Vinayagam, B. 2011. AFM surface investigation of Inconel 825 with multi wall carbon nano tube in electrical discharge machining process using Taguchi analysis. Archives of Civil and Mechanical Engineering, 11(1): 149-170.

Prabhu, S. and Vinayagam, B. 2013. AFM nano analysis of Inconel 825 with single wall carbon nano tube in die sinking EDM process using Taguchi analysis. Arabian Journal for Science and Engineering, 38(6): 1599-1613.

Pramanik, A., Basak, A., Islam, M.N. and Littlefair, G. 2015. Electrical discharge machining of 6061 aluminium alloy. Transactions of Nonferrous Metals Society of China, 25(9): 2866-2874.

Prasad, N.E. and Wanhill, R.J.H. 2017. Aerospace materials and material technologies. Retrieved from http://lib.myilibrary.com?id=970018

Purohit, R., Verma, C. and Shekhar, P. 2012. Electric discharge machining of 7075 Al-10 wt.% SiC_p composites using rotary tube brass electrodes. International Journal of Engineering Research and Applications, 2(2): 411-423.

Rahman, M., Khan, M.A.R., Kadirgama, K., Maleque, M. and Bakar, R.A. 2011. Parametric optimization in EDM of Ti-6Al-4V using copper tungsten electrode and positive polarity: A statistical approach. Mathematical Methods and Techniques in Engineering and Environmental Science, 1: 23-29.

Rajesha, S., Sharma, A. and Kumar, P. 2010. Some aspects of surface integrity study of electro discharge machined Inconel 718. Proceedings of the 36th International MATADOR Conference, 14-16 July 2010, University of Manchester, Manchester (UK).

Rajesha, S., Sharma, A. and Kumar, P. 2012. On electro discharge machining of Inconel 718 with hollow tool. Journal of Materials Engineering and Performance, 21(6): 882-891.

Rajurkar, K.P. and Yu, Z. 2000. 3D micro-EDM using CAD/CAM. CIRP Annal – Manufacturing Technology, 49(1): 127-130.

Rajyalakshmi, G. and Ramaiah, P.V. 2013. Multiple process parameter optimization of wire electrical discharge machining on Inconel 825 using Taguchi grey relational analysis. The International Journal of Advanced Manufacturing Technology, 69(5-8): 1249-1262.

Rao, P.S., Ramji, K. and Satyanarayana, B. 2014. Experimental investigation and optimization of wire EDM parameters for surface roughness, MRR and white layer in machining of aluminium alloy. Procedia Materials Science, 5: 2197-2206.

Roethel, F., Kosec, L. and Garbajs, V. 1976. Contribution to the micro-analysis of the spark eroded surfaces. Annals of the CIRP, 25(1): 135-140.

Sarkar, S., Mitra, S. and Bhattacharyya, B. 2005. Parametric analysis and optimization of wire electrical discharge machining of γ-titanium aluminide alloy. Journal of Materials Processing Technology, 159(3): 286-294.

Satishkumar, D., Kanthababu, M., Vajjiravelu, V., Anburaj, R., Sundarrajan, N.T. and Arul, H. 2011. Investigation of wire electrical discharge machining characteristics of Al6063/SiC_p composites. The International Journal of Advanced Manufacturing Technology, 56(9): 975-986.

Scott, D., Boyina, S. and Rajurkar, K.P. 1991. Analysis and optimization of parameter combinations in wire electrical discharge machining. International Journal of Production Research, 29(11): 2189-2207.

Selvakumar, G., Sornalatha, G., Sarkar, S. and Mitra, S. 2014. Experimental investigation and multi-objective optimization of wire electrical discharge machining (WEDM) of 5083 aluminum alloy. Transactions of Nonferrous Metals Society of China, 24(2): 373-379.

Singh, A. and Ghosh, A. 1999. A thermo-electric model of material removal during electric discharge machining. International Journal of Machine Tools and Manufacture, 39(4): 669-682.

Singh, P.N., Raghukandan, K., Rathinasabapathi, M. and Pai, B. 2004. Electric discharge machining of Al–10% SiCp as cast metal matrix composites. Journal of Materials Processing Technology, 155: 1653-1657.

Singh, P., Kumar, A., Beri, N. and Kumar, V. 2010a. Influence of electrical parameters in powder mixed electric discharge machining (PMEDM) of hastelloy. Journal of Engineering Research and Studies, 976: 7916.

Singh, P., Kumar, A., Beri, N. and Kumar, V. 2010b. Some experimental investigation on aluminum powder mixed EDM on machining performance of hastelloy steel. International Journal of Advanced Engineering Technology, 1: 28-45.

Singh, G., Sen, I., Gopinath, K. and Ramamurty, U. 2012. Influence of minor addition of boron on tensile and fatigue properties of wrought Ti-6Al-4V alloy. Materials Science and Engineering: A, 540: 142-151.

Singh, S. 2012. Optimization of machining characteristics in electric discharge machining of 6061Al/Al$_2$O$_3$p/20P composites by grey relational analysis. The International Journal of Advanced Manufacturing Technology, 63(9): 1191-1202.

Snoeys, R. and Van Dijck, F. 1971. Investigation of electro discharge machining operations by means of thermo-mathematical model. CIRP Annal – Manufacturing Technology, 20(1): 35-37.

Stráský, J., Janeček, M., Harcuba, P., Bukovina, M. and Wagner, L. 2011. The effect of microstructure on fatigue performance of Ti-6Al-4V alloy after EDM surface treatment for application in orthopaedics. Journal of the Mechanical Behavior of Biomedical Materials, 4(8): 1955-1962.

Thesiya, D., Rajurkar, A. and Patel, S. 2014. Heat affected zone and recast layer of Ti-6Al-4V alloy in the EDM process through scanning electron microscopy (SEM). Journal of Manufacturing Technology Research, 6: 41.

Tsai, H., Yan, B. and Huang, F. 2003. EDM performance of Cr/Cu-based composite electrodes. International Journal of Machine Tools and Manufacture, 43(3): 245-252.

Ulutan, D. and Ozel, T. 2011. Machining induced surface integrity in titanium and nickel alloys: A review. International Journal of Machine Tools and Manufacture, 51(3): 250-280.

Velmurugan, C., Subramanian, R., Thirugnanam, S. and Ananadavel, B. 2011. Experimental investigations on machining characteristics of Al 6061 hybrid metal matrix composites processed by electrical discharge machining. International Journal of Engineering, Science and Technology, 3(8): 87-101.

Wang, C.C. and Yan, B.H. 2000. Blind-hole drilling of Al$_2$O$_3$/6061Al composite using rotary electro-discharge machining. Journal of Materials Processing Technology, 102(1): 90-102.

Yadav, V., Jain, V.K. and Dixit, P.M. 2002. Thermal stresses due to electrical discharge machining. International Journal of Machine Tools and Manufacture, 42(8): 877-888.

Yan, B.H., Wang, C.C., Chow, H.M. and Lin, Y.C. 2000. Feasibility study of rotary electrical discharge machining with ball burnishing for Al$_2$O$_3$/6061Al composite. International Journal of Machine Tools and Manufacture, 40(10): 1403-1421.

Yan, B.H., Tsai, H.C., Huang, F.Y. and Lee, L.C. 2005a. Examination of wire electrical discharge machining of Al$_2$O$_3$/6061Al composites. International Journal of Machine Tools and Manufacture, 45(3): 251-259.

Yan, B.H., Tsai, H.C. and Huang, F.Y. 2005b. The effect in EDM of a dielectric of a urea solution in water on modifying the surface of titanium. International Journal of Machine Tools and Manufacture, 45(2): 194-200.

Yeo, S., Kurnia, W. and Tan, P. 2008. Critical assessment and numerical comparison of electro-thermal models in EDM. Journal of Materials Processing Technology, 203(1): 241-251.

Yilmaz, O. and Okka, M.A. 2010. Effect of single and multi-channel electrodes application on EDM fast hole drilling performance. The International Journal of Advanced Manufacturing Technology, 51(1-4): 185-194.

Yu, Z., Masuzawa, T. and Fujino, M. 1998. Micro-EDM for three-dimensional cavities- development of uniform wear method. CIRP Annal – Manufacturing Technology, 47(1): 169-172.

Yu, H.L., Luan, J.J., Li, J.Z., Zhang, Y.S., Yu, Z.Y. and Guo, D.M. 2010. A new electrode wear compensation method for improving performance in 3D micro-EDM milling. Journal of Micromechanics and Microengineering, 20(5): 055011.

Yuzawa, T., Magara, T., Imai, Y. and Sato, T. 1997. Micro EDM by thin rod electrode. Proceedings of Annual Assembly of Japan Society of Electrical Machining Engineers, 65-66.

PART III
Advanced Topics in SEM

Wire Spark Erosion Grinding

Andrew Rees

College of Engineering, Swansea University, Bay Campus, Crymlyn Burrows,
Swansea, SA1 8EN, United Kingdom
Email: andrew.rees@swansea.ac.uk

5.1 Introduction

Surface finish achievable by any machining process and its micro-versions
is an important aspect in determining its capabilities and optimizing its
parameters. A number of non-conventional and hybrid machining processes
which combine the capabilities of different complementary processes have
been developed to extend the capabilities of the existing micro-machining
processes. Wire spark erosion grinding (WSEG) is a hybrid machining process
(Masuzawa et al. 1985, Yu et al. 1998, Weng et al. 2003), developed to broaden
the application area of WSEM and the range of components manufactured by
it, by adding a rotary submergible spindle to the WSEM process for enabling
the machining of cylindrical components.

A number of researchers (Masuzawa et al. 1985, Yu et al. 1998, Masuzawa
et al. 2005) have investigated the evolution and other aspects of WSEG,
mainly focusing on the removal of a relatively small volume of material,
dressing electrodes from 150 µm down to 5 µm in diameter for use in drilling
and milling operations on die-sinking µ-SEM machines. Qu et al. (2002a,
2002b) studied the machining of free-form cylindrical parts by WSEG, with
an objective to extend the capabilities of WSEM by introducing an additional
rotary axis in its machine. They studied the effects of pulse on-time, rotational
speed, and wire feed rate on the surface integrity and roundness of macro-
sized manufactured parts; hence, their findings cannot be directly used in the
applications of WSEM for micro-sized parts. Juhr et al. (2004) highlighted the
importance of selecting correct parameters of WSEM process for the main
cutting operation, and concluded that material properties and surface finish
obtained during the main cut can be improved marginally by performing
follow-up finishing by WSEG. Therefore, special attention needs to be paid to
the surface finish obtained in the main cutting operation as it determines, to a
large extent, the achievable final surface roughness while machining of micro-

sized components by WSEG. Piltz and Uhlmann (2006) studied the effects of three different approaches for manufacturing cylindrical components by die-sinking μ-SEM and μ-WSEM, focusing on process behaviour in terms of pulse stability, hydrodynamic behaviour of dielectrics, machine dependent gaps, and feed controls. However, they did not study the effects of these process characteristics on the surface finish of the manufactured components and the optimisation of the μ-WSEM process. Rees et al. (2007) studied the influence of machining strategy on the quality of electrodes manufactured by WSEG. But these machining strategies cannot be used directly for combining WSEG with μ-WSEM, for the volume of material to be removed is generally much higher.

It can be concluded from the review of past research on manufacturing of cylindrical components by WSEM that (i) the prime focus has been mostly on the manufacturing of macro-sized components, (ii) effects of process parameters on material removal rate (MRR) (and not on surface roughness) have been studied using statistical methods and mathematical models (Mohammadi et al. 2008, Haddad and Fadei Tehrani et al. 2008, Matoorian et al. 2008), (iii) the exact mechanism of WSEG has not been investigated, and (iv) the development of dedicated machining strategies is required for combining WSEG with μ-WSEM.

Further development of WSEG needs work on its material removal mechanism, studying and analysing the effects of spindle speed, flushing pressure, pulse-off and pulse-on times and open circuit voltage on the surface roughness of the machined workpiece, and optimizing these important parameters. This chapter focuses on machining of micro-sized axis-symmetric components by WSEG, particularly on the achievable surface roughness and workpiece preparation to overcome the limitations of the existing machining strategies. It also details the study of pulse waveforms to quantify the fundamental differences between WSEG and μ-WSEM proceses, and development of models for on-the-machine prediction of surface roughness of the parts manufactured by WSEG, using inductive learning techniques to create a rule set.

5.2 Apparatus and Workpiece Preparation for WSEG

Apparatus for WSEG can be developed on a μ-WSEM machine by using a collet of the submergible rotating spindle on which a cylindrical workpiece can be mounted, with micro-wire as the cutting tool for machining micro-sized cylindrical workpieces (Fig. 5.1).

In an instance of preparing a micro-sized workpiece for WSEG, a 2 mm long workpiece was sliced from a piece of 3 mm diameter cut into a square cross-section of 100 μm side (by μ-WSEM) at four different angular positions with 90° offsets between them (Fig. 5.2). After this preparatory step, WSEG can be performed on the continuously rotating spindle. Initial attempts involved the WSEG of a 3 mm diameter rotating workpiece consisting of 94% tungsten carbide and 6% cobalt (produced by sintering powder with an average grain size of 0.3 µm) by µ-WSEM, using a 50 µm diameter brass coated steel wire; but the results were not satisfactory as the removal of a higher volume of material led to continuous wire breakage. To circumvent this issue, the workpiece was prepared by µ-WSEM, removing majority of the material in pre-WSEG operation.

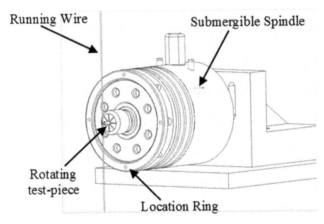

Fig. 5.1: Experimental apparatus for WSEG, developed on a µ-WSEM machine

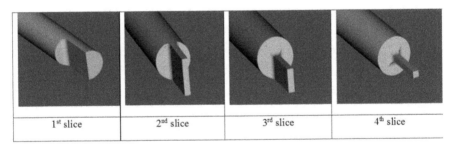

Fig. 5.2: Slicing of a 3000 µm diameter workpiece into a 100 µm square cross-section to prepare micro-sized workpieces for WSEG

The material allowance while preparing a workpiece for WSEG can be calculated using the following relation:

$$S_{profile} = D_1 + S_{pos} + S_g \qquad (1)$$

Here, D_1 is the target diameter after WSEG; S_{pos} is the run-out error in the part; and S_g is the spark gap (as shown in Fig. 5.3).

The result of machining by WSEG is shown in Fig. 5.4(a), while Fig. 5.4(b) shows the resulting 100 µm cross-section. It is apparent that the machining conditions change when the workpiece rotates, leading to a change in surface finish along the workpiece, that is, sides 'A' and 'B' in Fig. 5.4(a). The reason for this inconsistency is the varying amount of material that had to be removed along the square cross-section, runout error in the workpiece, and the spark gap required to have a stable WSEG.

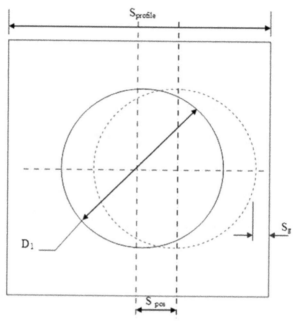

Fig. 5.3: Schematic for calculating material allowance during the preparation of a workpiece for WSEG

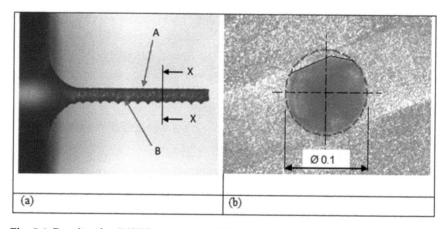

Fig. 5.4: Results after WSEG process: (a) 100 μm cross-section of the workpiece, and (b) cross-section at X-X

5.3 Experimental Investigations and Results

Experiments designed and conducted using Taguchi approach optimise the parameters of WSEG, with an objective of minimizing the resulting surface roughness. In one such experiment, each of five important parameters of the WSEG process (namely, rotational speed of the spindle, flushing pressure, pulse-off duration, open circuit voltage and pulse-on duration) were varied at three levels (Table 5.1). These parameters were selected because their values can be varied by the machine operator when performing the WSEG process, helping in on-the-machine process optimisation. Ranges and levels of each parameter were carefully selected to achieve a stable WSEG process (avoiding any condition which could lead to a wire breakage) and consistent results. Full factorial design of experiments with five parameters, each varying at three levels, would have required 243 experiments. Assuming that input parameters are independent, it is possible to use Taguchi L_{27} orthogonal array (Ross 1988, Phadke 1989), thus reducing the number of experiments to 27.

Table 5.1: Selected range of WSEG parameters available on the experimental apparatus and values selected for experimental investigations

Parameter (unit)	Range	Code	Level 1	Level 2	Level 3
Rotation speed of the spindle (RPM)	500 to 1500	A	500	1000	1500
Flushing pressure (MPa)	0 to 0.2	B	0	0.1	0.2
Pulse-off duration (µs)	12.5 to 42.5	C	12.5	27.5	42.5
Open circuit voltage (V)	100 to 200	D	100	150	200
Pulse-on duration (µs)	4.5 to 52.4	E	4.5	28.55	52.4

Four process parameters were continuously monitored during the experiments (Table 5.2). During machining, five data points per second for each parameter were recorded using a data acquisition system. This relatively low data acquisition rate was selected to capture the process details over longer time periods of machining and to make the volume of the acquired data manageable (the size of the formed data files over an average machining time of 3.5 minutes was 180 MB). The average surface roughness (R_a) of each workpiece machined by WSEG was measured using white light interferometer microscope, by analysing the scanned profiles along a sampling length of 250 µm and by applying a high-pass filter to remove their wavy characteristics. Subsequently, the collected data was pre-processed in 'LabVIEW'. The data was further processed using inductive learning techniques to create predictive models for on-the-machine assessment of the resulting surface finish by WSEG. For each experiment, average surface roughness values of every workpiece before and after WSEG was measured using white light interferometer microscope by analysing the scanned profiles along a sampling length of 250 µm and by applying a high-pass filter to remove their wavy characteristics. Table 5.3 presents 27 combinations of the input parameters used in the experiments, along with surface roughness values. To minimise the influence of possible stochastic factors on the resulting surface finish after WSEG, the experiments were carried out in a random order and due to time and cost constraints, no experiment was replicated, that is, only one experimental run was performed for each parametric combination.

Table 5.2: Monitoring parameters used in the experimentation of WSEG

Symbol (unit)	Description
V_x (mm/min)	Cutting speed along the x-axis
U_{fs} (V)	Average voltage between the wire electrode and workpiece
S (%)	Real value of servo-control parameter
P_w (kHz)	Pulse frequency

Table 5.3: Combinations of the input parameters for Taguchi's L_{27} orthogonal array and the corresponding values of average surface roughness

Run	A	B	C	D	E	Ra (µm)
1	1	1	1	1	1	1.32
2	1	1	1	1	2	1.09
3	1	1	1	1	3	0.64
4	1	2	2	2	1	1.66
5	1	2	2	2	2	1.70
6	1	2	2	2	3	2.12
7	1	3	3	3	1	1.27
8	1	3	3	3	2	1.58
9	1	3	3	3	3	1.70
10	2	1	2	3	1	1.09
11	2	1	2	3	2	3.20
12	2	1	2	3	3	2.09
13	2	2	3	1	1	1.34
14	2	2	3	1	2	0.89
15	2	2	3	1	3	1.14
16	2	3	1	2	1	1.02
17	2	3	1	2	2	0.80
18	2	3	1	2	3	1.54
19	3	1	3	2	1	1.05
20	3	1	3	2	2	1.17
21	3	1	3	2	3	1.04
22	3	2	1	3	1	0.87
23	3	2	1	3	2	0.96
24	3	2	1	3	3	1.18
25	3	3	2	1	1	1.51
26	3	3	2	1	2	1.05
27	3	3	2	1	3	0.93

To compare the surface roughness values produced by WSEG and µ-WSEM processes, tungsten carbide workpieces were produced using the parameters in Table 5.4, for performing one main and three trim cuts. Comparison of the two surface roughness profiles (Fig. 5.5) clearly shows that µ-WSEM gives an improved surface roughness, with $R_a = 0.20$ µm, compared to $R_a = 0.51$ µm yielded by WSEG. It can be concluded from these results that rotation of the workpiece in WSEG leads to significant differences in surface roughness, despite the same parameters. Therefore, the effects of the parameters of WSEG process on its performance were studied to improve surface roughness.

Table 5.4: Parameters used in µ-WSEM process

Parameter	Main cut	Trim cut 1	Trim cut 2	Trim cut 3
Flushing pressure (MPa)	1.2	1.2	1.2	1.2
Pulse-off duration	20 (µs)	500 (ns)	300 (ns)	200 (ns)
Open circuit voltage (V)	200	120	150	120
Pulse-on duration (µs)	40	2	2	2

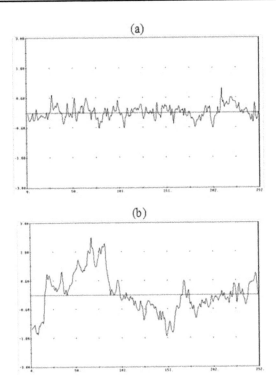

Fig. 5.5: Surface profiles produced by (a) µ-WSEM, and (b) WSEG

Analysis of variance (ANOVA) was done using the experimental results presented in Table 5.3 to evaluate the relative contribution of WSEG parameters on the surface roughness produced (Table 5.5). A confidence interval of 95% was used in these results, that is, the effect of any parameter having P-value less than 0.05 is considered to be significant. It can be seen in Table 5.5 that pulse-off duration has $P = 0.019$, implying that it has the highest significance on the resulting surface roughness, whereas open circuit voltage ($P = 0.137$) and spindle speed ($P = 0.161$) have lesser effect. The other two parameters – flushing pressure ($P = 0.792$) and pulse-on duration ($P = 0.743$) – have the least effect on the resulting surface roughness.

Table 5.5: Results of ANOVA for surface roughness results in Table 5.3

Source	D.F.	Seq SS	Adj SS	Adj MS	F-value	P-value
Rotation speed of the spindle (RPM)	2	0.8239	0.8239	0.4120	2.05	0.161
Flushing pressure	2	0.0950	0.0950	0.0475	0.24	0.792
Pulse-off duration	2	2.0612	2.0612	1.0306	5.13	0.019
Open circuit voltage	2	0.9045	0.9045	0.4523	2.25	0.137
Pulse-on duration	2	0.1216	0.1216	0.0608	0.30	0.743
Error	16	3.2134	3.2134	0.2008		
Total	26	7.2195				

Note: D.F.: degrees of freedom; Seq SS: sequential sum of squares; Adj SS: adjusted sum of squares; Adj MS: adjusted mean square

The normal probability plot (Fig. 5.6) was used to verify the assumption that the raw data had a normal distribution of errors as the plot in Fig. 5.6 resembles a straight line. The graphical relationship between the residuals

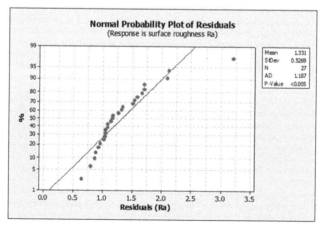

Fig. 5.6: Normal probability plot of residuals

and the fitted values (Fig. 5.7) shows that the experimental results satisfy two assumptions. In particular, there is no correlation between residuals and no error dependencies or violation of the constant variance assumption (Ross 1988). Only one outliers can be detected in Figs. 5.6 and 5.7, with the parametric combination: rotation speed of the spindle = 1000 RPM, flushing pressure = 0 MPa, pulse-off duration = 27.5 μs, open circuit voltage = 200 V, and pulse-on duration = 28.55 μs.

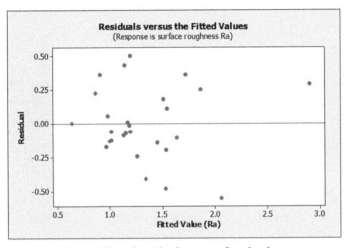

Fig. 5.7: Plot of residuals versus fitted values

5.4 Identification of Optimum Parameters for WSEG

In Taguchi design of experiment, signal-to-noise ratio (S/N ratio) is used as a measure of robustness to identify the controllable parameters of a process/ product, which reduce variability in the process/product by minimizing the effects of noise factors, that is, uncontrollable factors. Noise factors cannot be controlled during production or product use; however, they can be controlled during experimentation. The S/N ratio measures how responses vary with respect to their target values under different noise conditions. Maximizing the S/N ratio ensures a better quality process or product design. Taguchi design experiments use a two-step optimization process. The first step uses signal-to-noise ratio to identify the parameters which reduce variability. The second step identifies the parameters which move the mean value of the response closest to the target value, with no or minimal effect on the signal-to-noise ratio. Taguchi design experiments manipulate the noise factors, forcing variability to occur, and, from the results, the combination of optimum process/product parameters which make the process/product robust (resistant to variations caused by the noise factors) is identified. Higher values of signal-to-noise ratio (S/N) identify control factor settings that minimize the effect of noise factors

Taguchi loss function of smaller-the-better (SB) type was used to measure deviations in response (that is, the value of surface roughness) from its desired value (theoretically zero, which is practically impossible) and to identify the optimum values of the parameters. The loss function was further transformed into signal-to-noise (S/N) ratio, the high value of which indicates a better process response (the combination of optimum process parameters corresponds to the highest S/N ratio). Equation 2 presents the expression for signal-to-noise for smaller-the-better (SB) type loss function:

$$\left(\frac{S}{N}\right)_{SB} = -10\log\left(\frac{1}{n}\sum_{i=1}^{n} y_i^2\right) \tag{2}$$

Here, y_i is the surface roughness value for the ith experiment, and n is the number of experimental observations (Phadke 1989, Ross 1988). Table 5.6 presents the calculated values of S/N ratios for the three levels of WSEG process parameters.

Table 5.6: Signal-to-noise ratios for the different levels of the five parameters of WSEG

Level	Rotation speed of the spindle	Flushing pressure	Pulse-off duration	Open circuit voltage	Pulse-on duration
1	2.8311	2.0740	0.1268	0.5916	1.6829
2	2.4341	2.0088	4.0184	2.1834	1.9907
3	0.5972	1.7795	1.7172	3.0873	2.1887
Delta	2.2339	0.2945	3.8916	2.4957	0.5058
Rank	3	5	1	2	4

ANOVA was performed on these S/N ratios to study the significance of the five parameters considered for the WSEG process (Table 5.7). The results reveal that pulse-off duration, with P and F values of 0.019 and 5.13 respectively, has the highest influence on the process response, while open circuit voltage and spindle speed, with P values less than 0.05, have relatively lower effect. Flushing pressure and pulse-on duration, having P values more than 0.05, do not have significant effect on the performance of the WSEG process.

Table 5.7: Results of ANOVA for the S/N ratios of the WSEG process parameters

Parameter	D.F.	Seq SS	Adj SS	Adj MS	F	P	PCR
Rotation speed of the spindle	2	25.566	25.566	12.7831	1.90	0.181	11.3
Flushing pressure	2	0.431	0.431	0.2154	0.03	0.969	0
Pulse-off duration	2	68.907	68.907	34.4535	5.13	0.019	9.6
Open circuit voltage	2	28.738	28.738	14.3692	2.14	0.150	14.2
Pulse-on duration	2	1.170	1.170	0.5848	0.09	0.917	0
Error	16	107.511	107.511	6.7194			22.9
Total	26	232.323					

Figure 5.8 presents the main effects plots for S/N ratios of the five parameters considered for WSEG. From these graphs, it can be concluded that the 1st levels of rotational spindle speed and flushing pressure, 2nd level of pulse-off duration (or discharge interval) and 3rd levels of open circuit voltage and pulse-on-time duration constitute an optimum parametric combination, which yields the minimum value of S/N ratio. Table 5.8 presents the optimum values corresponding to the identified optimum level of these parameters. Since the identified optimum combination of WSEG parameters does not correspond to any of the parametric combinations of the 27 experimental runs presented in Table 5.3, one more experiment was conducted for confirming the results. The average surface roughness achieved with this combination was 0.57 μm. Its comparison with the results presented in Table 5.3 reveals that it is better than the average surface roughness values obtained for any parametric combination in the 27 experimental runs.

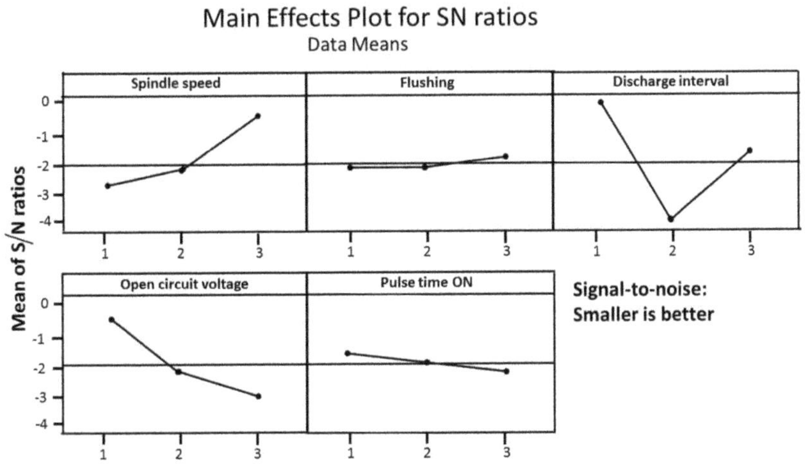

Fig. 5.8: Main effects plots for S/N ratios of WSEG parameters

Three subsequent trim cuts by μ-WSEM on the best WSEG-finished workpiece, using parameters mentioned in Table 5.4, further reduced the average surface roughness to 0.21 μm, comparable with the R_a value of 0.20 μm achieved with μ-WSEM (as mentioned earlier in this section). These results demonstrate that identification of optimum parameters in the WSEG process can yield a surface finish comparable to that produced by μ-WSEM.

The relatively high percentage contribution of errors as shown in Table 5.7 indicates the presence of interaction between process parameters. Figure 5.9 shows relatively strong interactions between the process parameters.

5.4.1 Comparison with μ-WSEM

Identification of optimum parameters of WSEG (Table 5.8) reveals that shorter pulse-on durations, lower open circuit voltage, and the resulting

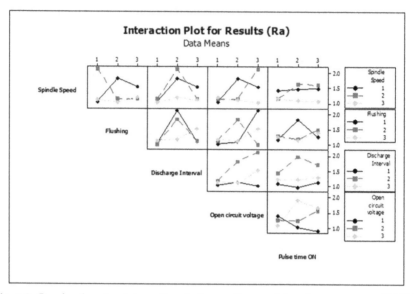

Fig. 5.9: Graphs showing interactions between the considered parameters of WSEG

low discharge energy help in achieving minimum surface roughness within the studied processing window. Such a conclusion seems obvious and also applicable to the μ-WSEM process. Performing an additional experiment was required to verify whether reduction in the discharge energy used in WSEG would also lead to surface roughness reduction in μ-WSEM. Figure 5.10 presents the results of this experiment, showing the surface roughness profiles of the workpieces machined by the μ-WSEM process – using the identified optimum parameters in Table 5.8 for WSEG, and the parameters in Table 5.4 for μ-WSEM. It is evident from these figures that reduction in discharge energy to the level required in WSEG results in negligible reduction in surface roughness by μ-WSEM, implying that reduction of discharge energy is only effective when the workpiece rotates, as in the case of WSEG. This is due to the fact that static discharge channels cannot be maintained over longer pulse-on durations, due to workpiece rotation. Discharge channels are interrupted above a particular rotational speed of

Table 5.8: Identified optimum values of WSEG parameters

Parameter	Level	Identified optimum value
Rotation speed of the spindle (RPM)	1	1500
Flushing pressure (MPa)	1	0.2
Pulse-off duration (μs)	2	12.5
Open circuit voltage (V)	3	100
Pulse-on duration (μs)	3	4.5

the workpiece, and discharge energy remains unchanged. This hypothesis is also supported by ANOVA results, suggesting that pulse-on duration has no statistically significant effect on surface roughness.

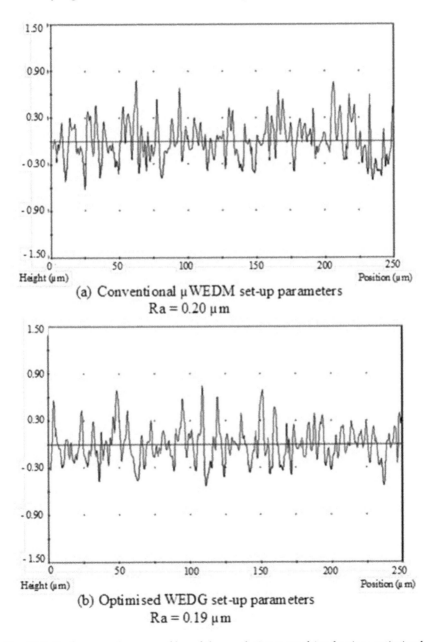

(a) Conventional μWEDM set-up parameters
$Ra = 0.20\ \mu m$

(b) Optimised WEDG set-up parameters
$Ra = 0.19\ \mu m$

Fig. 5.10: Surface roughness profiles of the workpieces machined using optimized parameters of (a) μWSEM and (b) WSEG

Figure 5.11 depicts distinct differences between the voltage profiles for µ-WSEM and WSEG processes. Open circuit voltages are typically much higher in µ-WSEM, whereas the actual values of pulse-on duration are higher for WSEG. An explanation for such a disparity in profiles can be sought in the process effects introduced by combining µ-WSEM and WSEG. Workpiece rotation leads to increased variations in the discharge channel gap. Its effect is an amplified retraction of the electrode wire, which can usually be done by sole adaptive servo-gap control of the machine. This leads to a faster and more efficient recovery of the working gap, thereby causing a reduction in the required open circuit voltage.

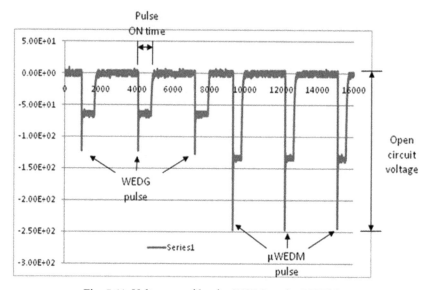

Fig. 5.11: Voltage profiles for WSEG and µ-WSEM

5.5 Inductive Model for Prediction of Surface Roughness

Information on WSEG process gathered through the monitoring of selected parameters can be used to gain an insight into the process and predict its performance. The acquired data can be pre-processed, using a suitable software such as LabVIEW, for help in generating predictive models for estimating the resulting surface roughness in WSEG, thus potentially minimising the necessary resources and time consumed in process optimisation, and identifying causal relationships between process parameters and the resulting surface quality. Though various machine learning techniques are available for developing predictive models, only a few attempts have been made to use such techniques in the area of SEM. Tsai and Wang (2001) applied neural networks to estimate the surface roughness of parts produced by SEM, as a function of workpiece material and electrode polarity, and compared the capabilities of various types of neural networks for predicting the resulting surface roughness. The drawback of using neural networks is that the created models represent a 'blackbox'; therefore, it is almost impossible for human experts to interpret them. Machine operators want to not only estimate the resulting surface roughness, but also understand the logic of the underlying model, thus allowing them to comprehend the relationship between process parameters and process performance.

In this context, the use of inductive learning algorithm in the form of rule sets, particularly RULES-F algorithm (Pham and Dimov 1997), to develop predictive models in the form of fuzzy rules can be readily interpreted by experts. The first step for successfully creating a predictive model is pre-processing of relevant data, and its selection and formatting into training examples for follow-up rule induction. The four WSEG parameters mentioned in Table 5.2 were used to generate predictive models by collecting five data points per second for each monitoring parameter during the experiments. Figure 5.12 illustrates the steady state of the WSEG process (along x-axis), which was used to gather training data. Mean and standard deviation (SD) of each monitoring parameter were used as input attributes to form the training set. Consequently, each data set for training purpose consisted of eight input parameters, and the corresponding average surface roughness value as the output (Table 5.9). When data is inadequate, it is not possible to apply typical model validation methods which require the available data to be split into training and testing sets. The same was applicable in this case; hence the entire data was used for training.

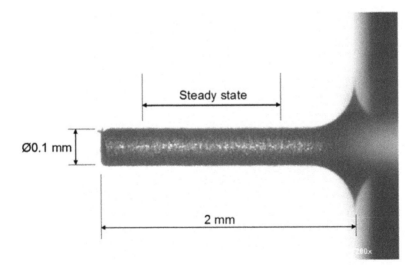

Fig. 5.12: Illustration of steady state machining by WSEG

Table 5.9: An example of mean and standard deviation values used as training data

Mean of V_x	SD of V_x	Mean of U_{fs}	SD of U_{fs}	Mean of Servo	SD of Servo	Mean of P_w	SD of P_w	R_a
0.54	37.765	20.12	11.317	24.547	6.654	29.232	3.803	1.32

Generation of rules with RULES-F requires the number of membership functions for output to be specified. To study the effect of this number on the performance of the generated rule sets, the algorithm was run 31 times by varying the number of membership functions from 2 to 32 during the training process. Each rule set was then used to predict the R_a values of the 32 training data sets. Subsequently, the absolute error between the predicted and the measured R_a values was computed for each data set. Figure 5.13 depicts the number of rules obtained for each number of fuzzy sets, along with their corresponding absolute mean error.

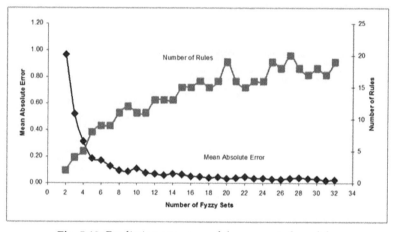

Fig. 5.13: Prediction accuracy of the generated models

Table 5.10 presents an example of a generic rule set generated using a triangular fuzzy set, represented by Tr (a, b, c), with 'ac' being its base and 'b' being the location of its apex (Fig. 5.14). The fuzzy set had only three membership values in terms of linguistic variables – low, medium and high.

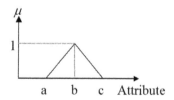

Fig. 5.14: Triangular fuzzy set used for developing a generic rule set

Accuracy of the generated rule set is relatively poor, with a mean absolute error of 0.52 (Fig. 5.15). However, this does not mean that the generated rules are inaccurate. In fact, each individual rule has been generated with 100% accuracy; but the output fuzzy sets used (low, medium, high) are too broad for accurate prediction during the defuzzification process. This suggests that

Table 5.10: Rule set generated using three values of the membership function for the fuzzy set

Examples covered	Fuzzy rules, including their interpretation by an expert
19	IF V_x Mean is Tr (11.25, 73.57, +∞) and U_{fs} mean is Tr (-∞, 0, 0.23) THEN R_a is Tr (0.64,1.92,3.2) IF [Mean of cutting speed along the x-axis] is medium high & [the mean of the voltage between the electrode wire and the work piece] is low THEN R_a is medium
17	IF U_{fs} SD is Tr (-∞, 2.18, 6.36) THEN R_a is Tr (0.64, 1.92, 3.2) IF [the SD of the voltage between the electrode wire and the work piece] is low THEN R_a is medium
13	IF V_x SD is Tr (-∞, 2.56, 11.83) and U_{fs} Mean is Tr (19.19, 30.48, +∞) THEN R_a is Tr(-0.64, 0.64, 1.92) IF [the SD of cutting speed along the x-axis] is low & [the mean of the voltage between the electrode wire and the work piece] is medium high THEN R_a is low
13	IF U_{fs} SD is Tr (6.64, 28.10, +∞) THEN Tr (-0.64, 0.64, 1.92) IF [the SD of the voltage between the electrode wire and the work piece] is medium high THEN R_a is low
1	IF V_x SD is Tr (3.43, 29.18, +∞) and V_x Mean is Tr (-∞, 10.25, 12.5) and P_w SD is Tr (-∞, 2.31, 2.43) THEN R_a is Tr (1.92, 3.2, 4.48) IF [the SD of cutting speed along the x-axis] is medium high, [Mean of cutting speed along the x-axis] is low and [the SD of the pulse frequency] is low THEN R_a is high

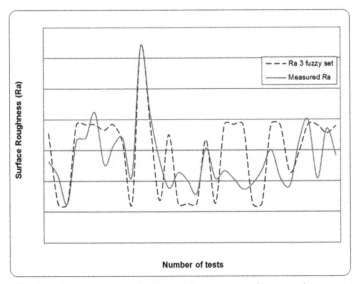

Fig. 5.15: Predicted and measured values of average surface roughness, using three values of membership function for the fuzzy set

there is a need to increase the number of fuzzy membership functions. It is also necessary for the generated rules to be meaningful and logical to gain some understanding of the underlining process behaviour. Analysis of the surface roughness produced by WSEG process shows that it is governed by interactions of the input variables; therefore, the rules generated using only three membership values for the fuzzy sets are not enough. Increase in the number of membership functions (N_f) by dividing a single linguistic term further (i.e. the 'low' linguistic term can be divided further into very low, low and relatively low) generally results in more accurate models by increasing the number of generated rules and formation of more specific rules. Hence, it becomes possible to identify some patterns in process behaviour and draw some logical interpretation of the process response. For example, $N_f = 9$ for the membership function shown in Fig. 5.16 allows the generation of nine rules, with a mean absolute error of only 0.08 (computed from the comparison of predicted and measured roughness as depicted in Fig. 5.17). Nevertheless, this also increases the complexity of the models, making them more difficult to interpret. Table 5.11 presents the rule set generated using the fuzzy set with nine membership functions. Unlike Table 5.10, the triangular fuzzy sets

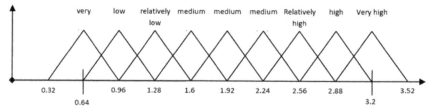

Fig. 5.16: Nine membership functions for the fuzzy set and linguistic terms used in them

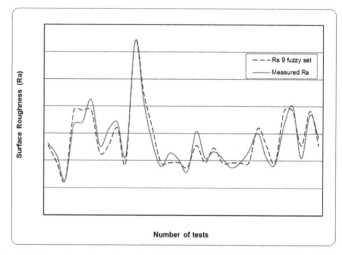

Fig. 5.17: Predicted and measured values of average surface roughness, using nine values of membership function for the fuzzy set

have been replaced by linguistic terms to facilitate the interpretation of the rules.

Table 5.11: Rule set generated using three values of membership function for the fuzzy set

Examples covered	Fuzzy Rules, including the interpretation by an expert
10	IF P_w SD is Tr (-∞, 2.31, 5.72), P_w Mean is Tr (26.08, 42.65, +∞) and U_{fs} SD is Tr (6.64, 28.1, +∞) THEN R_a is Tr (0.64, 0.96, 1.28) IF [the SD of the pulse frequency] *is medium low,* [the mean of the pulse frequency] *is medium high* and [the SD of the voltage between the electrode wire and the work piece] *is medium high THEN R_a is Low*
9	IF P_w SD is Tr (2.62, 4.10, 5.59), P_w Mean is Tr (27.28, 42.65, +∞), U_{fs} SD is Tr(2.18, 28.1, +∞) and SD of S is Tr(-∞, 2.46, 8.25) THEN R_a is Tr(0.96, 0.32, 1.6) IF [the SD of the pulse frequency] *is medium,* [the mean of the pulse frequency] *is medium high,* [the SD of the voltage between the electrode wire and the work piece] *is medium high* and [the SD of the real value of the servo control parameter] is low *THEN R_a is relatively Low*
5	IF P_w Mean is Tr (-∞, 26.04, 28.32) THEN R_a is Tr (1.6, 1.92, 2.24) IF [the mean of the pulse frequency] *is Low THEN R_a is medium*
5	IF V_x SD is Tr (2.57, 7.2, 11.83), V_x Mean is Tr (11.25, 73.58, +∞), P_w SD is Tr (-∞, 2.31, 4.02), P_w Mean is Tr (28.51, 42.65, +∞) and U_{fs} SD is Tr (2.18, 28.09, +∞) THEN R_a is Tr (0.64, 0.96, 1.28) IF [the SD of cutting speed along the x-axis] *is relatively low,* [Mean of cutting speed along the x-axis] *is medium high,* [the SD of the pulse frequency] *is medium low,* [the mean of the pulse frequency] *is medium high* and [the SD of the voltage between the electrode wire and the work piece] *is medium high THEN R_a is low*
4	IF V_x Mean is Tr (11.29, 73.58, +∞), P_w Mean is Tr (38.84, 42.65, +∞) and SD of S is Tr (-∞, 2.46, 9.57) THEN R_a is Tr (1.28, 1.6, 1.92) IF [Mean of cutting speed along the x-axis] *is medium high,* [the mean of the pulse frequency] *is high* and [the SD of the real value of the servo control parameter] *is low THEN R_a is medium low*
3	IF U_{fs} SD is Tr (-∞, 2.18, 11.42) and SD of S is Tr (9.57, 25.54, +∞) THEN R_a is Tr(0.32, 0.64 ,0.96) IF [the SD of the voltage between the electrode wire and the work piece] *is low* and [the SD of the real value of the servo control parameter] *is medium high THEN R_a is very low*
1	IF V_x SD is Tr(10.52, 29.18, +∞) and V_x Mean is Tr(-∞, 10.26, 33.41) THEN R_a is Tr(1.6, 1.92, 2.24) IF [the SD of cutting speed along the x-axis] *is medium high* and [Mean of cutting speed along the x-axis] *is low THEN R_a is medium*
1	IF V_x SD is Tr(3.43, 29.18, +∞), V_x Mean is Tr(-∞, 10.26, 12.50) and P_w SD is Tr(-∞, 2.31, 2.43) THEN R_a is Tr(2.88, 3.2, 3.52) IF [the SD of cutting speed along the x-axis] *is medium high,* [Mean of cutting speed along the x-axis] *is low* and [the SD of the pulse frequency] *is low THEN R_a is very high*

Generation of a rule set (Table 5.11) using a fuzzy set with higher number of membership functions significantly helps in interpreting the patterns which can be used to identify some generic trends in process behaviour. In addition to this, the following interesting rules associated with process settings leading to low and relatively low surface roughness were identified:

- Medium or medium low SD of 'P_w' resulted in low roughness in 26 examples
- Medium high mean of 'P_w' led to low roughness in 24 examples
- Medium high SD of 'U_{fs}' resulted in low roughness in 19 examples

Such generic patterns in process behaviour can be very useful in guiding machine operators in the identification of an optimum processing window for WSEG in order to minimize the resulting surface roughness. The rule sets generated using RULES-F can be used not only to estimate the resulting surface roughness, but also as a tool for analysing the process 'footprint', which is particularly useful while machining parts with varying cross-sectional areas.

5.6 Summary

Addition of a rotary submergible spindle to the μ-WSEM equipment helps in broadening its application area and allows machining of cylindrical components. The resulting process is referred to as WSEG. This chapter presented the effects of spindle speed, flushing pressure, pulse-off duration, open circuit voltage and pulse-on duration on the surface roughness resulting from WSEG. It described the method for workpiece preparation to overcome the limitations of the existing machining strategies. ANOVA of the signal-to-noise (S/N) ratio of the obtained results was used to identify the optimum parameters for the WSEG process. Inductive learning algorithms were used to generate generic rule sets to predict the resulting surface roughness after WSEG. Hence, the following concluding remarks can be made from this chapter:

- The machining strategy evolved to prepare a workpiece for WSEG minimises the effects of the inherent concentricity errors associated with the use of submergible spindles.
- A Taguchi DOE helped in assessing the statistical significance of pulse-off duration with regards to the resulting surface roughness after WSEG on a μ-WSEM setup.
- The identified optimum parameters of the WSEG process included the highest levels of spindle speed (1500 RPM) and flushing pressure (0.2 MPa), and the lowest levels of pulse-off duration (12.5 μs), open circuit voltage (100 V) and pulse-on duration (4.5 μs).
- Comparison between the shapes of voltage pulses of μ-WSEM and WSEG shows fundamental differences in their behaviours, which can be

explained with the combined effect of these two processes. In particular, rotation of the workpiece in WSEG leads to increased variations in the discharge channel gap, and an amplified retraction of the electrode wire. Ultimately, this reduces the open circuit voltage required for WSEG, due to a faster and more efficient recovery of the working gap.

- Reduction in the discharge energy of μ-WSEM improves the resulting surface roughness in the WSEG process negligibly.
- Generic rule sets developed using inductive learning algorithms are a simple and cost-effective method for on-the-machine prediction of surface roughness resulting from WSEG, making it possible to identify patterns, combinations of process parameters, and process 'footprint', which can lead to a low or relatively low surface roughness. This requires the SD of cutting speed along the x-axis to be medium or medium low, mean of cutting speed along the x-axis to be medium high, and the SD of pulse frequency to be medium high.

References

Haddad, M.J. and Fadaei Tehrani, A. 2008. Material removal rate (MRR) study in the cylindrical wire electrical discharge turning (CWEDT) process. Journal of Materials Processing Technology, 199(1-3): 369-378.

Juhr, H., Schulze, H.P., Wollenberg, G. and Künanz, K. 2004. Improved cemented carbide properties after wire-SEM by pulse shaping. Journal of Materials Processing Technology, 149(1-3): 178-183.

Masuzawa, T., Fujino, M. and Kobayashi, K. 1985. Wire electro-discharge grinding for micro-machining. CIRP Annals – Manufacturing Technology, 34(1): 431-434.

Masuzawa, T., Yamaguchi, M. and Fujino, M. 2005. Surface finishing of micropins produced by WSEG. CIRP Annals – Manufacturing Technology, 54(1): 171-174.

Matoorian, P., Sulaiman, S. and Ahmad, M. 2008. An experimental study for optimization of electrical discharge turning (EDT) process. Journal of Materials Processing Technology, 204(1-3): 350-356.

Mohammadi, A., Tehrani, A.F., Emanian, E. and Karimi, D. 2008. Statistical analysis of wire electrical discharge turning on material removal rate. Journal of Materials Processing Technology, 205(1-3): 283-289.

Phadke, M.S. 1989. Quality Engineering Using Robust Design. Prentice Hall, Englewood Cliffs, New Jersey, USA.

Pham, D.T. and Dimov, S.S. 1997. Efficient algorithm for automatic knowledge acquisition. Pattern Recognition, 30(7): 1137-1143.

Piltz, S. and Uhlmann, E. 2006. Manufacturing of Cylindrical Parts by Electrical Discharge Machining Processes. Proceedings of the 1st International Conference on Micromanufacturing (ICOMM), pp. 227-233.

Qu, J., Shih, A.J. and Scattergood, R.O. 2002a. Development of the cylindrical wire electrical discharge machining process, Part 1: Concept, design, and material removal rate. Transactions of the ASME: Journal of Manufacturing Science and Engineering, 124(3): 702-707.

Qu, J., Shih, A.J. and Scattergood, R.O. 2002b. Development of the cylindrical wire electrical discharge machining process, Part 2: Surface integrity and roundness. Transactions of the ASME: Journal of Manufacturing Science and Engineering, 124(3): 708-714.

Rees, A., Dimov, S., Ivanov, A., Herrero, A. and Uriarte, L. 2007. Micro-electrode discharge machining: Factors affecting the quality of electrodes produced on the machine through the process of wire electro-discharge machining. Proceedings of the Institution of Mechanical Engineers, Part B: Journal of Engineering Manufacture, 221: 409-418.

Ross, P.J. 1988. Taguchi Techniques for Quality Engineering. McGraw Hill, New York.

Tsai, K.M. and Wang, P.J. 2001. Predictions on surface finish in electrical discharge machining based upon neural network models. International Journal of Machine Tools and Manufacture, 41(10): 1385-1403.

Weng, F.T., Shyu, R.F. and Hsu, C.H. 2003. Fabrication of micro-electrodes by multi-SEM grinding process. Journal of Materials Processing Technology 140: 332-334.

Yu, Z.Y., Masuzawa, T. and Fujino, M. 1998. 3D Micro-SEM with simple shape electrode. International Journal of Electrical Machining, 3: 7-12.

Spark Erosion Based Hybrid Processes

Afzaal Ahmed[1*], M. Rahman[2] and A. Senthil Kumar[2]
[1] Department of Mechanical Engineering, IIT Palakkad, Kerala, India
[2] Department of Mechanical Engineering, NUS Singapore

Development of advanced engineering materials with exceptionally superior mechanical, chemical, thermal and physical properties has presented unprecedented challenges to the present manufacturing industries. This has motivated and allowed designers to come up with complex design requirements. Combination of these aforementioned problems restricts the application of conventional manufacturing processes, thus making their adoption inappropriate and uneconomical. This necessitates the development of novel hybrid/compound processes to manufacture products of desired quality. The basic concept behind any hybrid machining process is the judicious combination of two or more machining processes to overcome their individual limitations and achieve the desired material removal rate (MRR), with the targeted accuracy. However, according to the International Academy for Production Engineering (CIRP), a *hybrid machining process is a process based on the simultaneous and controlled interaction of process mechanisms and/or energy sources/tools having a significant effect on process performance* (Lauvers 2011).

Spark erosion machining (SEM), commonly known as electrical discharge machining (EDM), is a non-contact thermal energy based process in which material is removed by melting and vaporization caused due to the heat generated by a sequence of repetitive sparks in the presence of a dielectric medium (Kunieda et al. 2005). It is one of the most widely used non-conventional processes to machine electrically conductive materials. However, MRR and surface integrity given by SEM frequently fail to meet the growing demands of shorter lead time and flawless surface quality for the present manufacturing industries. Moreover, complicated 3D surfaces, tight tolerances and extreme design constraints have led machining researchers to explore possible combinations of SEM with other existing processes to machine these advanced materials economically and accurately. This

*Corresponding author: afzaalahmed86@gmail.com

chapter summarizes the development of SEM-based hybrid processes aimed at overcoming the aforesaid issues, detailing their working principles and mechanisms of material removal.

6.1 Types of Spark Erosion Based Hybrid Processes

SE-based hybrid processes can be broadly classified into the following two groups (Fig. 6.1):

- **Compound/combined processes:** Both the constituent machining processes are responsible for material removal from the workpiece in a compound/combined hybrid process. The two different material removal mechanisms of the constituent processes work simultaneously to enhance the overall performance of the combined process. For example, in a compound process of SEM and electrochemical machining (ECM), both SEM and ECM are individually responsible for controlling material removal from the work material.
- **Assisted processes:** In such processes, the primary machining process is responsible for the removal of material, whereas the role of the secondary process is to assist the primary process. For instance, in ultrasonic-assisted SEM (UASEM), the workpiece material is removed by SEM (through melting and vaporization), whereas ultrasonic vibrations facilitate the removal of debris from the narrow machining gap.

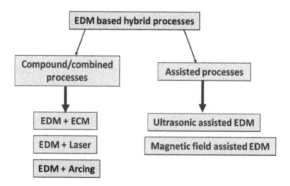

Fig. 6.1: Classification of spark erosion based hybrid processes

6.2 Hybrid Machining Processes of SEM and ECM

In SEM, every discharge of spark over the workpiece and the tool surface removes a very small volume of material by melting and vaporization. Ejection of the molten material creates a crater on the workpiece and the tool electrode surface. The size of these craters depends on the discharge energy of the spark (Yu et al. 2003). The spark erosion machined (SEMed) surface has multiple overlapping discharge craters (Lee et al. 2003, Ekmekci et al. 2005,

Kurnia et al. 2009). Therefore, an SEMed surface usually has a high surface roughness.

A typical SEMed surface has micro-cracks, migrated material from the tool, and white layer formation (Lee and Tai 2003, Ekmekci 2007, Ahmed 2016). ECM is also a non-contact type non-conventional machining process. In this process, material is removed by electrolytic dissolution of electrically conducting anodic workpiece in the presence of a suitable electrotype, producing an approximately complimentary shape of the cathodic tool on the workpiece when the required value of voltage between them is applied through a direct current (DC) supply unit (McGeough 1974). Consequently, the surface generated by an ECM process is usually very smooth, with negligible defects (Bhattacharyya et al. 2004). The performance of ECM does not depend on the mechanical properties of the workpiece, thus enabling the machining of very hard, tough and high strength materials even by tools made of a soft and low strength material. Nonetheless, there are certain limitations to the ECM process: (i) generation of oxygen at the anode, leading to the oxidation of the workpiece, making it electrically non-conducting if a proper type of electrolyte is not selected and/or its required flow not maintained; (ii) sensitivity towards the properties of electrolytes (which depend on its temperature, which increases with the square of the current flowing); (iii) difficulty in the ECM of alloys, due to different electrochemical equivalent and thermal properties of its constituent process; and (iv) chemical damage to the ECMed surface (such as inter-granular attack, selective dissolution, etching, composition alternation of an alloy, etc.). Therefore, an appropriate combination of SEM and ECM processes mitigates their adverse effects and exploits their capabilities and advantages simultaneously. Examination of the feasibility of this combination requires experimentation at micro-level. SEM and ECM can be associated in two different ways to develop hybrid process: (i) sequential SEM and ECM, and (ii) simultaneous spark erosion electrochemical machining (SSECM).

6.2.1 Sequential SEM and ECM

The objective of sequential combination of SEM and ECM is to improve the poor surface integrity and to remove the thermally damaged zones created by the SEM process. Hung et al. (2006) used 85% concentration H_3PO_4 as the electrolyte in sequential μ-ECM of SEM-drilled micro-holes. They used low voltage in the range of 1 to 5 V during the experiments due to the very high conductivity of the electrolyte. They observed a reduction in the taper of the micro-hole, micro-burrs over the machined wall surface, and surface roughness (from 2.11 to 0.69 μm), using a voltage of 2 V for a machining duration of 5 minutes. Skoczypiec and Rusraj (2014) developed and studied sequential SEM and ECM on a single machine tool, focusing on the manufacture of micro-parts. Figure 6.2 shows the photograph of the prototype of the machine developed and used. Application of sequential

ECM and SEM almost halved the machining time in comparison with spark erosion micro-machining or electric discharge micro-machining (EDMM) (Fig. 6.3). Zeng et al. (2012) used milling for the preparation of the workpiece, and then used micro-SEM and micro-ECM sequentially (using the same

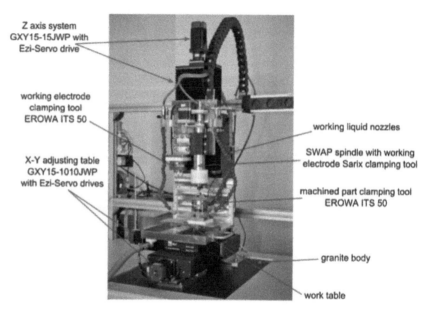

Fig. 6.2: Photograph of the machine prototype for sequential SEM and ECM (Skoczypiec and Rusraj 2014 © Elsevier. Reprinted with permission)

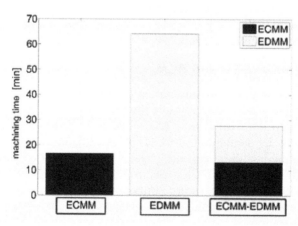

Fig. 6.3: Comparison of machining times of electrochemical micro-machining (ECMM), electro discharge micro-machining (EDMM), and sequential combination of ECMM and EDMM (Skoczypiec and Rusraj 2014 © Elsevier. Reprinted with permission)

electrode), as depicted in Fig. 6.4, to manufacture 3D features; however, they used different dielectric fluids. They observed that this combination resulted in a better surface quality. The surface roughness value reduced from 0.707 µm to 0.143 µm, with the complete removal of recast layer, burrs, micro-pores and craters (Fig. 6.5). Though dimensional accuracy in ECM is difficult to control due to stray corrosion at the edges of the machined feature, this combined process provides a better control over dimensional accuracy by using suitable machining conditions and appropriate tool path.

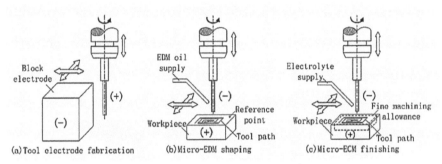

Fig. 6.4: Schematic of sequential micro-SEM and micro-ECM (Zeng et al. 2012 © Elsevier. Reprinted with permission)

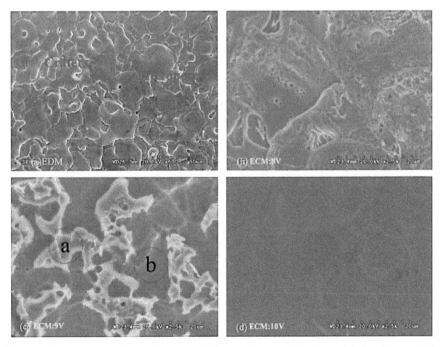

Fig. 6.5: Improvement in surface morphology after combined machining (Zeng et al. 2012 © Elsevier. Reprinted with permission)

Takahata et al. (1997) used micro-SEM and the hybrid of micro-ECM and lapping as a final finishing step to enhance the surface integrity of the machined surface, with an objective to take advantage of the dissolution effect of ECM and the polishing effect of the colloidal aqueous electrolyte solution containing Al_2O_3 abrasive grains. The electrolytic solution is responsible for electrochemical dissolution, whereas the impact caused by the abrasive grains provides the necessary polishing action. The required kinetic energy for the abrasive grains to move in the machining gap is provided by the rotating electrode. It was reported that a mirror-like surface was obtained with a maximum surface roughness of 32 nm, after machining for 120 seconds. Kurita and Hattori (2006) used a similar approach to finish hardened steel (61 HRC) after micro-SEM, using a copper electrode and Al_2O_3 abrasive powder of 2-13 µm size mixed with the electrolyte. They found that ECM-lapping gave lower roughness than that by SEM (or EDM), ECM or polishing alone (Fig. 6.6).

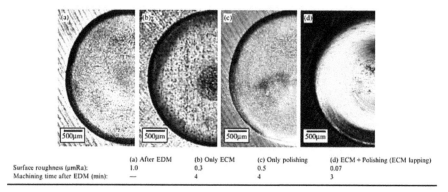

	(a) After EDM	(b) Only ECM	(c) Only polishing	(d) ECM + Polishing (ECM lapping)
Surface roughness (µmRa):	1.0	0.3	0.5	0.07
Machining time after EDM (min):	—	4	4	3

Fig. 6.6: Surface roughness obtained by different machining processes (Kurita and Hattori 2006 © Elsevier. Reprinted with permission)

6.2.2 Simultaneous Spark Erosion Electrochemical Machining (SSECM)

Deionized water is the most widely used alternative dielectric fluid in SEM, after hydrocarbon oils. It is considered as an eco-friendly fluid for thermal machining processes as it does not produce harmful gases like CO, CH_4, etc. Higher MRR and low electrode wear rate (EWR) are additional special features that make deionized water a better choice than hydrocarbon oil (Abbas et al. 2007). Many studies have been conducted to evaluate the performance of deionized water in SEM, pioneered by Jeswani (1981). He found that using distilled water as the dielectric, along with a pulse energy of 72-288 mJ, resulted in a higher MRR, lower EWR and better surface finish, but poor dimensional accuracy, compared to that with kerosene used as dielectric. This is due to the dissolution of the workpiece material caused by electrochemical action. Kim and Chu (2007) conducted micro-spark erosion milling using deionized water as the dielectric and found that the low resistivity of deionized water caused electrochemical dissolution, thereby increasing the machining gap and reducing tool wear. They also reported that the amount of electrochemical dissolution can be significantly reduced by increasing the resistivity of deionized water. Nguyen et al. (2013) reported that the initial gap between the tool electrode and the workpiece also defines the rate of electrochemical dissolution and the latter is significantly reduced with increase in the initial gap after attaining its peak value at 5 µm gap.

Nguyen et al. (2012a) developed a hybrid machining process called simultaneous spark erosion chemical machining (SSECM), combining micro-SEM and micro-ECM, using the abovementioned concept of machining in low-resistivity deionized water. This simultaneous action in one hybrid process resulted in an enhanced surface finish and dimensional accuracy (Nguyen et al. 2012b). Figure 6.7 presents the working principle of SSECM. This process uses short voltage pulses supplied by a resistance-capacitance (RC) type power supply. In the beginning, these short voltage pulses are applied across the tool and the workpiece electrode. Then, the electrode moves closer to the workpiece (Fig. 6.7a). As soon as the machining gap reaches a critical value, breakdown of deionized water takes place and a spark is produced (Fig. 6.7b). Subsequently, material removal occurs by melting and vaporization due to the thermal energy of sparks, and the workpiece surface gets covered with multiple discharge craters (Fig. 6.7b). Furthermore, the machining gap increases due to material removal by electrical discharges. However, owing to the low conductivity of deionized water, electrochemical dissolution begins as shown in Fig. 6.7c. Short voltage pulses assure that the dissolution is localized. This ECM action dissolves the depressions created by the sparks, thereby significantly improving the surface roughness of the workpiece. Further, the electrode is lowered again and the entire process is repeated (Fig. 6.7d). It was reported that in order to give sufficient time for the electrochemical dissolution to take place, the feed rate must be low. Figure 5.8 illustrates the sample machined by SSECM process. Simultaneous spark erosion chemical drilling (SSECD) and simultaneous spark erosion chemical milling (SSECMl) have also been attempted with an objective of producing micro-features with high machining accuracy and fine surface finish (Nguyen et al. 2012b).

6.2.2.1 Simultaneous Spark Erosion Chemical Drilling (SSECD)

Nguyen et al. (2012b) investigated the SSECD process to study the impact of feed rate, applying a 60 V DC voltage, and found that the wall of the micro-hole is covered with overlapping discharge craters for a feed rate 10 µm/s (as shown in Fig. 6.9a), whereas a feed rate of 0.2 µm/s resulted in a micro-hole with smooth and crater-free walls (Fig. 6.9b). This can be explained on the basis of the fact that a 10 µm/s feed rate is rather fast, which does not allow much time for electrochemical dissolution to happen. On the other hand, a feed rate of 0.2 µm/s provides ample time for electrochemical reaction to take

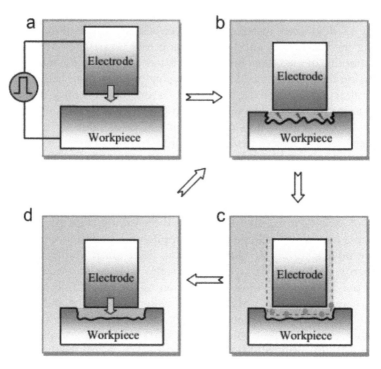

Fig. 6.7: Working principle of simultaneous spark erosion chemical machining (SSECM) (Nguyen et al. 2012a © Elsevier. Reprinted with permission)

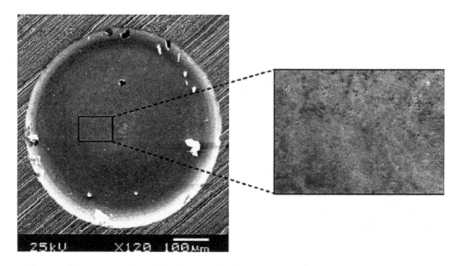

Fig. 6.8: Surface machined by SSECM (voltage = 80 V, feed rate = 0.2 µm/s, stray capacitance = 7 pF) (Nguyen et al. 2012b © Elsevier. Reprinted with permission)

place on the cylindrical surface of the machined micro-hole. Furthermore, use of a continuous DC voltage supply causes a violent electrochemical reaction resulting in excessive material removal. Consequently, the machined shape gets distorted, compromising the machining accuracy. Therefore, short voltage pulses are utilized instead of continuous DC to confine the area of material dissolution. Figure 6.10 shows the SEM micrograph of the micro-hole fabricated with short voltage pulses of 500 kHz frequency, 30% duty cycle and 0.2 μm/s feed rate, in low resistivity deionized water. The surface obtained showed no signs of overlapping craters and there is no significant distortion of the micro-hole. Therefore, it can be concluded that SSECD is capable of producing micro-holes with good surface finish and high dimensional accuracy.

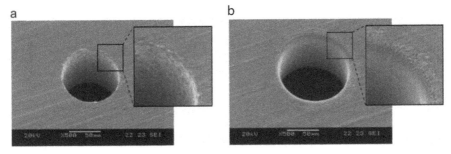

Fig. 6.9: Micro-holes machined by SSECD (with 60 V DC supply) at different feed rates: (a) 10 μm/s and (b) 0.2 μm/s (Nguyen et al. 2012b © Elsevier. Reprinted with permission)

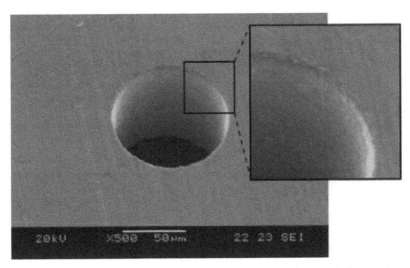

Fig. 6.10: Micro-hole fabricated by SSECD (with 500 kHz pulses, 30% duty ratio and 0.2 mm/s feed rate) (Nguyen et al. 2012b © Elsevier. Reprinted with permission)

6.2.2.2 Simultaneous Spark Erosion Chemical Milling (SSECMl)

In addition to SSECD, SSECMl experiments have been conducted to fabricate precise 3D features for micro-applications. This process is similar to SSECD, except that the electrode moves downwards by a specific layer depth to retain the original shape of the electrode. Figure 6.11 illustrates the working principle of this process. Machining speed, a function of layer thickness and horizontal scanning feed rate, has to be adjusted for obtaining effective SSECMl (Nguyen et al. 2012b). Figure 6.12 shows micro-slots of 5 µm depth for different feed rates in the range of 50 to 10 mm/s. The depth of each layer was 0.2 um at the applied voltage 60 V with a pulse frequency of 500 kHz and 30% duty ratio.

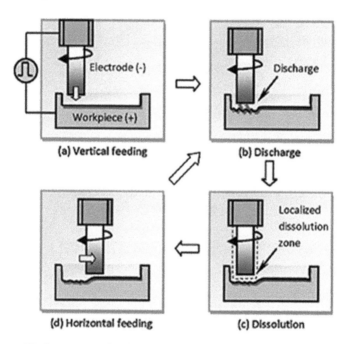

Fig. 6.11: Working principle of simultaneous spark erosion electrochemical milling (Nguyen et al. 2012b © Elsevier. Reprinted with permission)

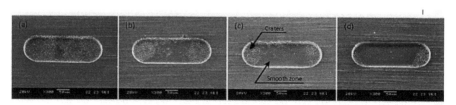

Fig. 6.12: SSECMl fabricated Micro-slots using short voltage pulses at different scanning feed rates: (a) 50 mm/s, (b) 30 mm/s, (c) 20 mm/s, and (d) 10 mm/s (Nguyen et al. 2012b © Elsevier. Reprinted with permission)

It has been observed that discharge craters were present over the machined surface at 50 mm/s scanning feed rate. These overlapping craters however start vanishing with reduction in the scanning feed rate, due to an increase in the duration of electrochemical dissolution; consequently, very few craters are observed at 10 mm/s. Layer depth is another important parameter in SSECMl. Figure 6.13 depicts the micro-slots machined at 10 µm/s scanning feed rate, with a higher layer depth of 0.5 µm and 1 µm respectively. It can be noted that discharge craters start to appear with an increase in layer depth. This is due to an increase in the amount of material to be removed in every feed. Consequently, a prolonged discharge occurs, causing material removal by melting, thus reducing the machining time required for dissolution. This is evident from Fig. 6.13b, wherein the machined micro-slot surface is completely covered with overlapping craters when the layer depth was selected as 1 µm. Figure 6.14 presents images of micro-features with various shapes. It is clear from these images that the machined surface is completely covered with discharge craters at 50 µm/s scanning feed rate, whereas it becomes smoother at 10 µm/s scanning feed rate.

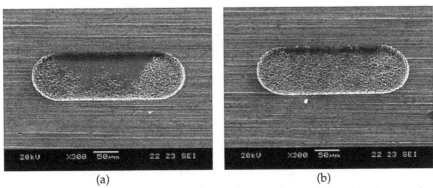

(a)   (b)

Fig. 6.13: SSECMl-machined micro-slots at different layer depths: (a) 0.5 µm and (b) 1 µm (Nguyen et al. 2012b © Elsevier. Reprinted with permission)

Surface morphology of SSECMl machined surfaces has been shown is Fig. 6.15. Discharge craters of diameters 3-4 µm can be seen at 50 mm/s feed rate (Fig. 6.15a), whereas a feed rate of 10 mm/s yields a smooth surface with no signs of craters (Fig. 6.15b). Measurement of surface roughness profile shows that a 50 mm/s feed rate results in average surface roughness of 142 nm, which reduces to 22 nm for a feed rate of 10 mm/s. This shows that SEDCMl yields a better surface finish due to the effect of electrochemical dissolution. Similarly, improvements in dimensional accuracy have also been observed (Fig. 6.16).

The discussions in the previous paragraphs clearly demonstrate that SSECMl is capable of yielding a good surface finish and enhanced dimensional accuracy. This makes it suitable for micro-molding applications in the fabrication of micro-features. Figure 6.17 shows the comparison of micro-features fabricated by spark erosion milling and SSECMl.

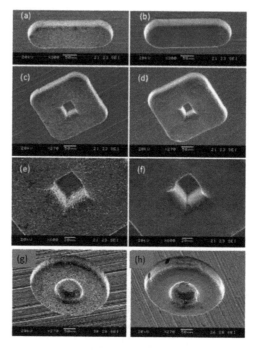

Fig. 6.14: Micrographs of SSECMl-machined micro-slots and micro-features at different feed rates: (a, c, e, g) 50 mm/s and (b, d, f, h) 10 mm/s (Nguyen et al. 2012b © Elsevier. Reprinted with permission)

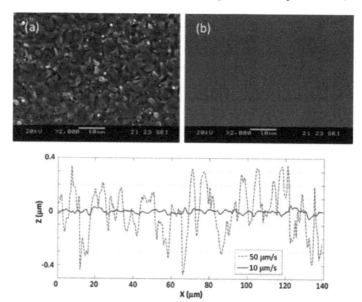

Fig. 6.15: Topography and profiles of surfaces generated at different feed rates: (a) 50 mm/s and (b) 10 mm/s (Nguyen et al. 2013 © Elsevier. Reprinted with permission)

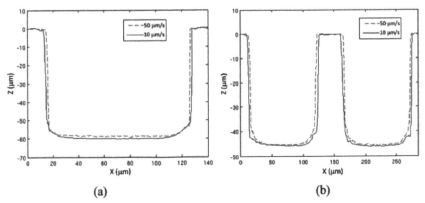

Fig. 6.16: Profiles of micro-slots fabricated at different feed rates (Nguyen et al. 2013 © Elsevier. Reprinted with permission)

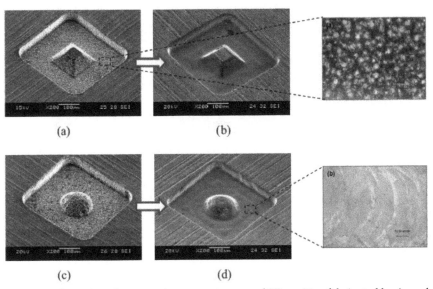

(a)   (b)

(c)   (d)

Fig. 6.17: Scanning electron microscope images of 3D cavities fabricated by (a and c) spark erosion milling and (b and d) SSECMl (Nguyen et al. 2013 © Elsevier. Reprinted with permission)

6.3 Hybrid Processes of SEM and Laser Machining

Laser machining yields a higher MRR, but it produces a larger recast layer and HAZ as well (Dausinger 2000). Whereas SEM gives a good surface quality, it is slower than laser machining. Therefore, a combination of SEM followed by laser machining is often adopted. Li et al. (2006) investigated the finishing of micro-holes made by laser beam drilling (LBD) followed by spark erosion drilling (SED), for manufacturing next generation fuel injection nozzles. They found that this hybridization overcame the problems of lower productivity of SED, and the formation of a large recast layer and HAZ associated with LBD. Figure 6.18 compares the micro-hole made by LBD, SED (146 µm diameter), and LBD followed by SED (140 µm diameter). The combination of LBD and SED reduced drilling time by 70% and cost by 42%, and increased the productivity by 90%, without compromising the quality of the micro-hole.

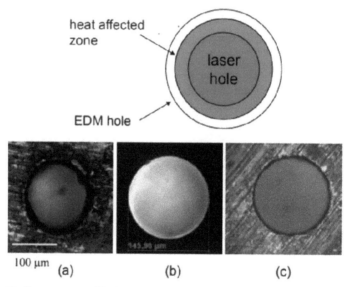

Fig. 6.18: Comparison of holes made by (a) laser beam drilling (LBD), (b) spark erosion drilling (SED), and (c) LBD followed by SED (Li et al. 2006 © Elsevier. Reprinted with permission)

Al-Ahmari et al. (2016) used a combination of laser ablation and SED for drilling holes in a 500 µm thick sheet of Ni-Ti based shape memory alloy and found that it resulted in a 40-65% increase in MRR. Fleischer et al. (2006) also used a combination of LBD and SED to manufacture wear-resistant mold inserts in hardened steel for powder injection molding on a 5-axis machining center (Fig. 6.19) having the capabilities of die-sinking SEM and ultra-short pulse Nd:YAG laser (15 ps pulse interval) of 355 nm wavelength. The laser was focused on the workpiece up to a diameter of 7 µm with the help of an objective. They used pico-second Nd-YAG laser of 20 mW power and 40% pulse overlap to fabricate internal micro-gears with a depth of 50 µm and width of tooth as 20 µm (Fig. 6.20a). The same internal micro-gear was manufactured by drilling holes through SEM and subsequent laser ablation to cut the teeth (Fig. 6.20(b)). Comparing the micrographs in Figs. 6.20a and 6.20b, it can be clearly observed that the combination of laser ablation and SEM produces an internal micro-gear of much better quality, with its teeth having better edge definition, compared to that manufactured by laser ablation alone.

Fig. 6.19: Machining center combining SEM and laser ablation (Fleischer et al. 2006 © Springer. Reprinted with permission)

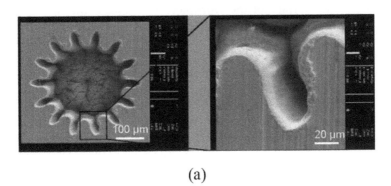

(a)

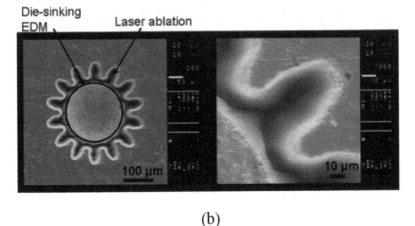

(b)

Fig. 6.20: Micro-gear made of hardened steel, manufactured by: (a) laser ablation using 355 nm wavelength pico-second Nd:YAG laser and (b) a combination of die-sinking SEM and laser ablation (Fleischer et al. 2006 © Springer. Reprinted with permission)

6.4 Hybrid Spark Erosion Arc Machining (HSEAM)

Though SEM is one of the most widely used non-conventional machining processes, the MRR of this process is very low (Abbas et al. 2007). Subsequently, SEM has been successfully hybridized with arc machining to enhance the existing capabilities of SEM, with the resultant process named as hybrid spark erosion and arc machining (HSEAM) or hybrid electric discharge arc machining (HEDAM) (Ahmed 2016 and Ahmed et al. 2017). This combination addresses the problems associated with fast machining of difficult-to-machine materials such as nickel-based superalloys, titanium alloys, etc.

6.4.1 Working Principle

Figure 6.21 shows the experimental apparatus for HSEAM. The tool electrode descends to the surface of the workpiece to decrease the machining gap. Once the critical gap is achieved, the dielectric fluid ionizes and creates a plasma channel for the discharge of arcs to begin, which elevates the thermal energy density. This leads to a quick melting of the workpiece, leaving behind molten debris and a crater on the machined surface. Thus, the mechanism of material removal in HSEAM can be summarized as a process of quick thermal melting of a workpiece, followed by the removal of the molten debris with high pressure dielectric fluid flushing. HSEAM can be operated in three modes: (i) **Mode 1:** when the voltage across the DC power supply is kept lower than the voltage across the pulsed power-supply ($V_{DC} < V_{PDC}$), (ii) **Mode 2:** when DC and pulsed power-supply voltages are equal ($V_{DC} = V_{PDC}$), and (iii) **Mode 3:** when the voltage across the DC power supply is higher than that of the pulsed power-supply ($V_{DC} > V_{PDC}$).

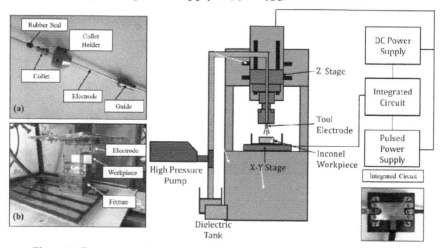

Fig. 6.21: Experimental apparatus for HSEAM: (a) collet-electrode-guide arrangement and (b) electrode-workpiece arrangement (Ahmed 2016)

Mode 1 ($V_{DC} < V_{PDC}$): The working principle of HSEAM for this mode can be explained with the help of Fig. 6.22. In the absence of current flow from the DC power supply, the material is machined only during the t_{on} time, which is exactly the same as SEM. When current is supplied from a DC power source to the electrode-workpiece system, material removal occurs during the t_{off} time as well. As shown in Fig. 6.22a, during the t_{on} period, diode D2 operates in forward bias mode, while D1 functions in reverse bias mode, as the voltage across the DC power supply is lower than the pulse power supply. At this instant, current is drawn only from the pulsed power supply. On the other hand, during t_{off} period, diode D1 is forward biased, while D2 is reverse biased as the power supply voltage across the pulsed power supply is zero. As a result, at this instant, the system draws current only from the DC source. Initially, when the power supply is turned on, the servo controller automatically moves the electrode close to the workpiece. This movement allows the electric field to attain its maximum strength when the electrode and the workpiece are closest to each other. Additionally, a low resistivity dielectric fluid initiates the plasma channel with a single spark. Generally, in the case of normal SEM, during the t_{off} period, current is no longer supplied to the system, resulting in the collapse of the plasma channel. On the contrary, in HSEAM, the plasma channel is sustained as the system is able to draw current from the DC power source during the t_{off} period as well. However, owing to the rotation of the electrode and a high flushing pressure, the plasma channel would eventually deionize, causing a new cycle to begin.

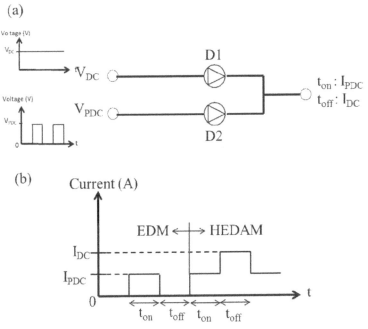

Fig. 6.22: Working principle of HSEAM in mode 1 ($V_{DC} < V_{PDC}$) (Ahmed 2016)

Mode 2 ($V_{DC} = V_{PDC}$): In this mode, as shown in Fig. 6.23, both the diodes D1 and D2 are forward biased during the t_{on} period, since the voltage across them is the same. Hence, current supplied to the system is proportionate to the summation of both the power supplies during t_{on}. However, during the t_{off} period of pulsed supply, the system draws current only from the DC source. It is difficult to operate the system in this working mode as the two voltages cannot be maintained at the same magnitude with respect to voltage drop across the diodes.

Mode 3 ($V_{DC} > V_{PDC}$): Here, since voltage across the constant DC supply is higher than the pulsed power supply, diode D1 is always forward biased. Therefore, regardless of t_{on} and t_{off}, the system draws current only from the DC power supply, thereby resulting in continuous arcing and making the pulsed power supply redundant. Continuous arcing at high temperatures melts the workpiece violently and non-uniformly, which is detrimental to the process of machining. Moreover, large quantities of debris generated in this mode may result in ineffective flushing.

In short, during HSEAM, SEM works as a catalyst to enable the formation of the initial plasma channel by sparking. Following the initial spark, conductivity of the plasma channel gradually reduces; but it is still adequate

(a)

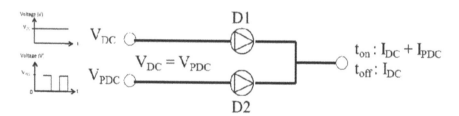

(b)

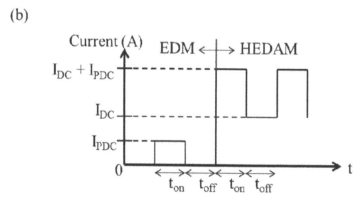

Fig. 6.23: Working principle of HSEAM in mode 2 ($V_{DC} = V_{PDC}$) (Ahmed 2016)

for the arc to sustain. However, the plasma channel eventually deionizes and collapses due to the rotation of the electrode and a high flushing pressure, thereby triggering a new cycle.

6.4.2 Comparative Study of SEM and HSEAM

MRR is the volume of workpiece material removed per unit time, whereas electrode wear ratio (EWR) is the percentage of volumetric electrode wear with respect to MRR. Figure 6.24 shows a comparison of MRR and EWR for drilling holes in Inconel 718 (of two different depths) by HSEAM and SEM. In this case, the dielectric flushing pressure was set at 8 MPa, DC current at 50A, and the electrode tool rotated at 2000 rpm during drilling by SEM. It has been reported that machining current significantly influences the volume and size of debris formed in SEM (Tanjilul et al. 2018); therefore, a high pressure of 8 MPa was required to flush the debris from the machining zone. Another experiment can be conducted without using flushing of dielectric to understand the significance of high-pressure flushing of dielectric fluid in HSEAM. Figure 6.25 shows the SEM images after performing the HSEAM process on the workpiece surface for four seconds, with an objective to study the mechanism of material removal and the process of debris formation.

The following observation can be made from Figs. 6.24 and 6.25:

- Drilling by SEM process results in higher EWR than MRR, whereas drilling by HSEAM process yields a higher MRR than EWR, for both

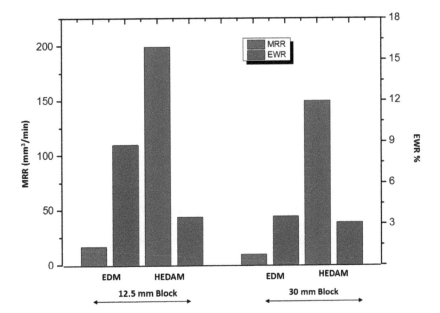

Fig. 6.24. Comparison of MRR and EWR of hole drilling in Inconel 718 by SEM (or EDM) and HSEAM (or HEDAM) for two values of hole depth (Ahmed 2016)

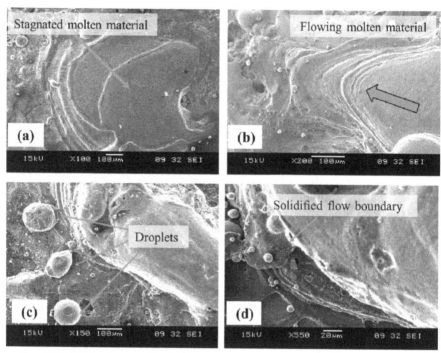

Fig. 6.25: SEM micrographs showing material removal mechanism and molten debris flow: (a) stagnated molten debris when no flushing of dielectric is used in SEM, (b) flowing molten debris with dielectric flushing, (c) resolidified spherical droplets, and (d) solidified crater boundary (Ahmed 2016)

thicknesses of Inconel 718. MRR for drilling the 12.5 mm thick Inconel 718 piece by HSEAM process is 200 mm³/min, almost 12 times than that obtained by SEM; whereas EWR in HSEAM is about 3.5%, which is much lower than that in drilling by SEM.

- MRR and EWR decrease with increase in hole depth in Inconel 718 for both SEM and HSEAM processes. This is attributed to the increase in ineffectiveness of dielectric flushing as the hole depth increases, which causes the debris to get trapped and remain at the bottom of the hole. The debris trapped in the gap at the bottom causes frequent electrode retractions, while machining and secondary discharges reduce the cutting speed. However, HSEAM has demonstrated relatively better results in terms of MRR, even at a hole depth of 30 mm.
- HSEAM process drilled Inconel 718 at a depth of 30 mm in 3 mins 40 seconds, which is much faster than conventional SEM. It also exhibited better performance in terms of energy efficiency, compared to conventional SEM (Ahmed et al. 2017).
- Without flushing, the molten debris gets stagnated and resolidified over the surface of the workpiece (Fig. 6.25a). Flushing of dielectric starts the

flow of the molten debris (Fig. 6.25b). An interesting observation is that spherical droplets of molten debris are deposited on the workpiece, as a result of fusion of high pressure flushing and the arcs (Fig. 6.25c). The end of flow of the molten debris as a solidified boundary is illustrated in Fig. 6.25d.

6.5 Vibration-assisted SEM

Major effects of using high frequency or ultrasonic vibrations in SEM include: (a) formation of an alternating pressure wave of the dielectric fluid, and (b) cavitation and pumping action in the inter-electrode gap (IEG). This facilitates circulation of the dielectric and removal of the debris from the narrow IEG. Consequently, effectiveness of arc discharges improves, leading to more efficient machining (Endo et al. 2008, Tong et al. 2008). There are two ways of vibration transfer in SEM:

1. Vibration to the tool electrode
2. Vibration to the workpiece

6.5.1 Providing Vibrations to the Tool Electrode

Kremer et al. (1989) made one of the earliest attempts to provide ultrasonic vibrations to the tool electrode. Their equipment consists of a piezoelectric transducer connected to a cross-shaped sonotrode which transmits and amplifies the horizontal ultrasonic vibrations received from the transducer in the vertical direction. The tool electrode has two parts brazed together: a metallic part to transmit the ultrasonic vibrations, and a graphite electrode for performing the SEM process. Ultrasonic vibrations generate high frequency alternate pressure variation. A decrease in the machining gap increases the pressure, leading to a low MRR, whereas an increase in the machining gap decreases the pressure, resulting in a higher MRR. Furthermore, it is also observed that the heat affected layer reduces and shows minimum micro-cracks. Kremer et al. (1991) conducted experimental investigations to understand the relationship between pressure in the machining gap and the effect of discharge. They also proposed a theoretical model to describe the phenomenon in this process. Srivastava and Pandey (2012) studied the process of ultrasonic-assisted cryogenically cooled SEM (UACSEM) on M2 HSS workpiece material, using the equipment shown in Fig. 6.26. Cryogenics provide the necessary cooling to the electrode, thus reducing heat accumulation in the tool electrode, and, in turn, EWR (Figs. 6.27a and 6.27b). External vibrations create turbulence due to continuous variation of pressure in the machining gap. This allows less liquid material to re-solidify over the machined surface, resulting in improved surface roughness (Figs. 6.27c and 6.27d). The use of this process to machine micro-channels of size 25 μm, in titanium alloy, reduces surface roughness by 19-43%. Moreover, high

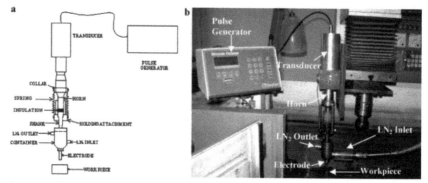

Fig. 6.26: Ultrasonic-assisted cryogenically cooled SEM (UACSEM): (a) schematic diagram of the vibration assembly and (b) compound assembly on SEM machine (Srivastava and Pandey 2012 © Elsevier. Reprinted with permission)

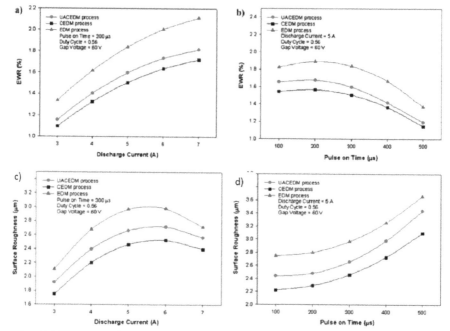

Fig. 6.27: Variation of (a) EWR with discharge current, (b) EWR with pulse-on time, (c) surface roughness with discharge current, and (d) surface roughness with pulse-on time (Srivastava and Pandey 2012 © Elsevier. Reprinted with permission)

frequency vibration generates pumping action, which helps in more efficient removal of debris from the machining zone, thus increasing MRR.

6.5.2 Providing Vibrations to the Workpiece

Providing vibrations to the tool electrode creates instability in the machining system, which is unfavorable for achieving higher machining efficiency and accuracy. This problem becomes even more critical when the electrodes are very thin. Gao and Liu (2003) suggested providing high-frequency vibrations to the workpiece to overcome this problem. Thus, an ultrasonic transducer vibrates the workpiece instead of the tool electrode. This eliminates the requirement of a bulky tool electrode system by eliminating set of transducer, acoustic horn, and cone. Ultimately, this makes the whole apparatus much simpler and compact than ultrasonic-assisted SEM in which vibrations are provided to the tool. Figure 6.28 shows the apparatus for this process, along with the configuration of ultrasonic transducer and the working principle of the generator. The transducer is made up of four cylindrical piezoelectric elements, one mounted on top of another, according to their identical polarity. This piezoelectric transducer is mounted on the worktable of the micro-SEM system. The ultrasonic generator has the ability to adjust frequency, voltage and pulse duration of the vibration signals.

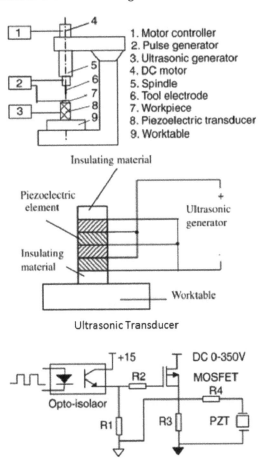

1. Motor controller
2. Pulse generator
3. Ultrasonic generator
4. DC motor
5. Spindle
6. Tool electrode
7. Workpiece
8. Piezoelectric transducer
9. Worktable

Insulating material

Piezoelectric element

Insulating material

Ultrasonic generator

Worktable

Ultrasonic Transducer

+15   DC 0-350V

R2   MOSFET
         R4

Opto-isolaor

R1   R3   PZT

Fig. 6.28: Apparatus for ultrasonic-assisted SEM in which vibrations are provided to the workpiece (Gao and Liu 2003 © Elsevier. Reprinted with permission)

The results of experiments conducted by Gao and Liu (2003) have shown that providing high frequency vibrations to the workpiece significantly improves the MRR of micro-SEM. They reported eight times higher MRR in ultrasonic vibration assisted micro-SEM of 0.5 mm thick stainless steel workpiece with a 43 µm diameter tool electrode than that in micro-SEM. Jahan et al. (2012) evaluated the effectiveness of providing low frequency vibrations to the workpiece during deep micro-hole drilling in tungsten carbide by micro-SED, and discovered that MRR increases and EWR decreases significantly with vibrations provided to the workpiece. Improvement is more pronounced at lower discharge energies, which in turn reduces the machining gap. This can be explained by the fact that providing vibrations to the workpiece helps the fresh dielectric to flood the machining gap, aiding in a more effective removal of debris, reduction in arcing and short-circuiting, and thus leading to an overall improvement in machining performance. Surface quality and dimensional accuracy improve due to significant reduction in arcing and short-circuiting. Garn et al. (2011) reaffirmed the fact that the total duration of arcing is reduced with application of vibrations to the workpiece in vibration-assisted micro-SEM as vibrations interrupt the generation of arcs by ending the arcing state. Figure 6.29 shows a comparison of the durations of arcing in micro-SEM and vibration-assisted micro-SEM. However, the number of arc discharges increases with increase in vibration frequency. Chern and Chuang (2006) concluded in their study that a larger feed and better surface finish are achievable by vibration-assisted micro-SEM.

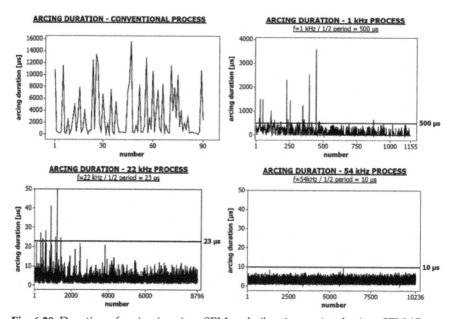

Fig. 6.29: Duration of arcing in micro-SEM and vibration-assisted micro-SEM (Garn et al. 2011 © Elsevier. Reprinted with permission)

6.6 Magnetic Field Assisted SEM

Removal of products of machining (debris) from the machining gap is one of the major challenges in SEM. Performance indicating parameters of SEM, such as process stability and the quality of the machined/finished surface obtained, are directly related to the ability of expelling debris from the machining zone. Complete and fast evacuation of debris from the narrow

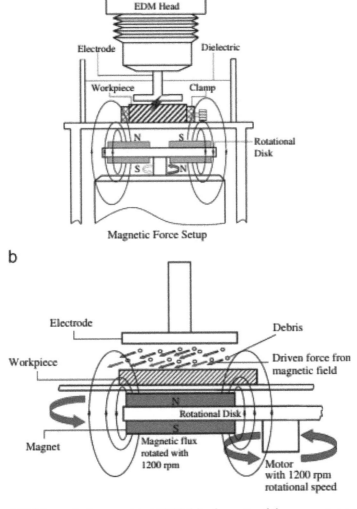

Fig. 6.30: Magnetic force assisted SEM: (a) schematic of the apparatus and (b) schematic diagram of the debris driven from the IEG due to magnetic force (Lin and Lee 2008 © Elsevier. Reprinted with permission)

IEG or machining gap is the defining factor in the improvement of machining efficiency of SEM. Several researchers have suggested different methods to solve this problem (Rajurkar and Pandit 1986, Masuzawa et al. 1992, Soni 1994). As discussed in the previous section, ultrasonic vibration is one of the techniques to enhance the removal of debris from the narrow IEG (Soni 1994, Thoe et al. 1999, Zhang et al. 2002). Lin and Lee (2008) suggested the use of magnetic force for effective removal of debris from the IEG in SEM and named this process as magnetic force assisted SEM (MFA-SEM). Figure 6.30 presents the apparatus designed by them for MFA-SEM. It consists of two magnets (0.3 T magnetic flux density) arranged symmetrically and connected to circular discs of 130 mm diameter and rotating at 1200 RPM, driven by an electrical motor. Fig. 6.30(b) depicts the movement of the debris being driven away by magnetic force during this process. An oscilloscope is used to exhibit the benefits of magnetic force assisted SEM over conventional SEM.

Lin and Lee (2008) compared the waveforms of discharges in MFA-SEM and SEM after 35 minutes of machining and observed that the number of effective discharges obtained with MFA-SEM is much higher than SEM, due to the accumulation of debris in the machining gap, leading to arcing and, consequently, instability of the process. This also enables MFA-SEM to result in a higher MRR (Lin et al. 2009, Teimouri and Baseri 2012). Joshi et al. (2011) investigated the use of pulsating magnetic field instead of continuous magnetic field in dry SEM, with an objective to enhance plasma ionization and control its expansion. They observed that confinement of plasma channel results in a higher thermal energy density, which enhances the melting of the workpiece, leading to a higher MRR. Furthermore, there is significant improvement in dimensional accuracy and surface quality as well. Also, the pulsating nature of the magnetic field reduces flux losses, compared to a continuous magnetic field. Lin et al. (2014) combined magnetic field with ultrasonic vibrations to perform a hybrid SEM of SKD 61 tool steel. They reported an increase in MRR, good surface finish and stable SEM performance. Debris evacuation rate also improved due to the combined action of ultrasonic vibrations and magnetic force.

6.7 Summary

This chapter presented details of various hybrid or compound processes developed using SEM as the base process, with the objective of overcoming its limitations and improving its performance. It covered hybrid processes developed by sequential and simultaneous combination of SEM with ECM, hybridized process of LBD followed by SED, hybrid process of spark and arc machining, vibration-assisted SEM, and magnetic field assisted SEM. Descriptions in this chapter provide insights into and guidelines for the development of more compound processes of SEM in future.

References

Abbas, N.M., Solomon, D.G. and Bahari, M.F. 2007. A review on current research trends in electrical discharge machining (EDM). International Journal of Machine Tools and Manufacture, 47(7): 1214-1228.

Ahmed, A., Rahman, A. and Kumar, A.S. 2017. A novel approach in deep hole drilling of Inconel 718. Proceedings of 7th International Conference of Asian Society for Precision Engineering and Nanotechnology (ASPEN). Seoul, South Korea, 14-17 Nov. 2017.

Ahmed, A. 2016. A Study on the Compound EDM-arcing and Deep Hole Drilling of Inconel 718. PhD Thesis, Department of Mechanical Engineering, National University of Singapore (NUS), Singapore.

Ahmed, A., Fardin, A., Tanjilul, M., Wong, Y., Rahman, M. and Kumar, A.S. 2018. A comparative study on the modelling of EDM and hybrid electrical discharge and arc machining considering latent heat and temperature-dependent properties of Inconel 718. The International Journal of Advanced Manufacturing Technology, 94(7): 2729-2737.

Al-Ahmari, A., Rasheed, M.S., Mohammed, M.K. and Saleh, T. 2016. A hybrid machining process combining micro-EDM and laser beam machining of Nickel–Titanium-based shape memory alloy. Materials and Manufacturing Processes, 31(4): 447-455.

Bhattacharyya, B., Munda, J. and Malapati, M. 2004. Advancement in electrochemical micro-machining. International Journal of Machine Tools and Manufacture, 44(15): 1577-1589.

Chern, G.L. and Chuang, Y. 2006. Study on vibration-EDM and mass punching of micro-holes. Journal of Materials Processing Technology, 180(1): 151-160.

Dausinger, F. 2000. Precise drilling with short-pulsed lasers. Proceedings of International Society for Optics and Photonics (SPIE) on 'High Power Lasers in Manufacturing' X. Chen, T. Fujioka and A. Matsunawa (Eds.). DOI: 10.1117/12.377015

Ekmekci, B. 2007. Residual stresses and white layer in electric discharge machining (EDM). Applied Surface Science, 253(23): 9234-9240.

Ekmekci, B., Elkoca, O. and Erden, A. 2005. A comparative study on the surface integrity of plastic mold steel due to electric discharge machining. Metallurgical and Materials Transactions B, 36(1): 117-124.

Endo, T., Tsujimoto, T. and Mitsui, K. 2008. Study of vibration-assisted micro-EDM—the effect of vibration on machining time and stability of discharge. Precision Engineering, 32(4): 269-277.

Fleischer, J., Schmidt, J. and Haupt, S. 2006. Combination of electric discharge machining and laser ablation in microstructuring of hardened steels. Microsystem Technologies, 12(7): 697-701.

Gao, C. and Liu, Z. 2003. A study of ultrasonically aided micro-electrical-discharge machining by the application of workpiece vibration. Journal of Materials Processing Technology, 139(1): 226-228.

Garn, R., Schubert, A. and Zeidler, H. 2011. Analysis of the effect of vibrations on the micro-EDM process at the workpiece surface. Precision Engineering, 35(2): 364-368.

Hung, J.C., Yan, B.H., Liu, H.S. and Chow, H.M. 2006. Micro-hole machining using micro-EDM combined with electropolishing. Journal of Micromechanics and Microengineering, 16(8): 1480.

Jahan, M., Wong, Y. and Rahman, M. 2012. Evaluation of the effectiveness of low frequency workpiece vibration in deep-hole micro-EDM drilling of tungsten carbide. Journal of Manufacturing Processes, 14(3): 343-359.

Jeswani, M. 1981. Electrical discharge machining in distilled water. Wear, 72(1): 81-88.

Joshi, S., Govindan, P., Malshe, A. and Rajurkar, K.P. 2011. Experimental characterization of dry EDM performed in a pulsating magnetic field. CIRP Annals – Manufacturing Technology, 60(1): 239-242.

Kim, B.H. and Chu, C.N. 2007. Micro electrical discharge milling using deionized water as a dielectric fluid. Journal of Micromechanics and Microengineering, 17(5): 867.

Kremer, D., Lebrun, J., Hosari, B. and Moisan, A. 1989. Effects of ultrasonic vibrations on the performances in EDM. CIRP Annals – Manufacturing Technology, 38(1): 199-202.

Kremer, D., Lhiaubet, C. and Moisan, A. 1991. A study of the effect of synchronizing ultrasonic vibrations with pulses in EDM. CIRP Annals – Manufacturing Technology, 40(1): 211-214.

Kunieda, M., Lauwers, B., Rajurkar, K.P. and Schumacher, B. 2005. Advancing EDM through fundamental insight into the process. CIRP Annals – Manufacturing Technology, 54(2): 64-87.

Kurita, T. and Hattori, M. 2006. A study of EDM and ECM/ECM-lapping complex machining technology. International Journal of Machine Tools and Manufacture, 46(14): 1804-1810.

Kurnia, W., Tan, P., Yeo, S. and Tan, Q. 2009. Surface roughness model for micro electrical discharge machining. Proceedings of the Institution of Mechanical Engineers, Part B: Journal of Engineering Manufacture, 223(3): 279-287.

Lauvers, B. 2011. Surface Integrity in Hybrid Manufacturing Processes. Procedia Engineering, 19: 241-251.

Lee, H.T., Rehbach, W.P., Tai, T.Y. and Hsu, F.C. 2003. Surface integrity in micro-hole drilling using micro-electro discharge machining. Materials Transactions, 44(12): 2718-2722.

Lee, H.T. and Tai, T.Y. 2003. Relationship between EDM parameters and surface crack formation. Journal of Materials Processing Technology, 142(3): 676-683.

Li, L., Diver, C., Atkinson, J., Giedl-Wagner, R. and Helml, H. 2006. Sequential laser and EDM micro-drilling for next generation fuel injection nozzle manufacture. CIRP Annals – Manufacturing Technology, 55(1): 179-182.

Lin, Y.C., Chen, Y.F., Wang, D.A. and Lee, H.S. 2009. Optimization of machining parameters in magnetic force assisted EDM based on Taguchi method. Journal of Materials Processing Technology, 209(7): 3374-3383.

Lin, Y.C., Chuang, F.P., Wang, A.C. and Chow, H.M. 2014. Machining characteristics of hybrid EDM with ultrasonic vibration and assisted magnetic force. International Journal of Precision Engineering and Manufacturing, 15(6): 1143-1149.

Lin, Y.C. and Lee, H.S. 2008. Machining characteristics of magnetic force-assisted EDM. International Journal of Machine Tools and Manufacture, 48(11): 1179-1186.

Masuzawa, T., Cui, X. and Taniguchi, N. 1992. Improved jet flushing for EDM. CIRP Annals – Manufacturing Technology, 41(1): 239-242.

McGeough, J.A. 1974. Principles of Electrochemical Machining. CRC Press, Boca Raton, Florida, USA.

Nguyen, M.D., Rahman, M. and Wong, Y.S. 2012a. Simultaneous micro-EDM and micro-ECM in low-resistivity deionized water. International Journal of Machine Tools and Manufacture, 54: 55-65.

Nguyen, M.D., Rahman, M. and Wong, Y.S. 2012b. Enhanced surface integrity and dimensional accuracy by simultaneous micro-ED/EC milling. CIRP Annals – Manufacturing Technology, 61(1): 191-194.

Nguyen, M.D., Rahman, M. and Wong, Y.S. 2013. Modeling of radial gap formed by material dissolution in simultaneous micro-EDM and micro-ECM drilling using deionized water. International Journal of Machine Tools and Manufacture, 66: 95-101.

Rajurkar, K.P. and Pandit, S. 1986. Formation and ejection of EDM debris. Transactions of ASME: Journal of Engineering for Industry, 108(1): 22-26.

Skoczypiec, S. and Ruszaj. A. 2014. A sequential electrochemical-electrodischarge process for micro-part manufactureing. Precision Engineering, 38(3): 680-690.

Soni, J. (1994). Microanalysis of debris formed during rotary EDM of titanium alloy (Ti 6A1 4V) and die steel (T 215 Cr12). Wear, 177(1): 71-79.

Srivastava, V. and Pandey, P.M. 2012. Effect of process parameters on the performance of EDM process with ultrasonic assisted cryogenically cooled electrode. Journal of Manufacturing Processes, 14(3): 393-402.

Takahata, K., Aoki, S. and Sato, T. 1997. Fine surface finishing method for 3-dimensional micro structures. IEICE Transactions on Electronics, 80(2): 291-296.

Tanjilul, M., Ahmed, A., Kumar, A.S. and Rahman, M. 2018. A study on EDM debris particle size and flushing mechanism for efficient debris removal in EDM-drilling of Inconel 718. Journal of Material Processing Technology, 255: 263-274.

Teimouri, R. and Baseri, H. 2012. Effects of magnetic field and rotary tool on EDM performance. Journal of Manufacturing Processes, 14(3): 316-322.

Thoe, T., Aspinwall, D. and Killey, N. 1999. Combined ultrasonic and electrical discharge machining of ceramic coated nickel alloy. Journal of Materials Processing Technology, 92: 323-328.

Tong, H., Li, Y. and Wang, Y. 2008. Experimental research on vibration assisted EDM of micro-structures with non-circular cross-section. Journal of Materials Processing Technology, 208(1): 289-298.

Yu, Z., Rajurkar, K.P. and Narasimhan, J. 2003. Effect of machining parameters on machining performance of micro-EDM and surface integrity. Proceedings of Annual ASPE Meeting, Portland, Orlando, USA.

Zeng, Z., Wang, Y. Wang, Z., Shan, D. and He, X. 2012. A study of micro-EDM and micro-ECM combined milling for 3D metallic micro-structures. Precision Engineering, 36(3): 500-509.

Zhang, Q., Zhang, J., Deng, J., Qin, Y. and Niu, Z. (2002). Ultrasonic vibration electrical discharge machining in gas. Journal of Materials Processing Technology, 129(1): 135-138.

Vibroacoustic Diagnostics of Spark Erosion Based Processes

Artur N. Porvatov[1]* and Michail P. Kozochkin[2]

[1] Assistant Professor, Electrical Engineering, Electronics and Automation
[2] Professor, High-effective Machining Technologies Moscow State Technological
University Stankin, Moscow, Russia

7.1 Introduction

Spark erosion (SE) based machining processes such as spark erosion machining (SEM) and wire SEM (WSEM) are irreplaceable for manufacturing high-accuracy parts with complicated geometries and internal cooling systems for different aerospace applications. At present, these processes are used in the manufacture of aviation parts from materials with superior mechanical properties but poor machinability (such as ceramics, high-temperature nano-materials based on oxide ceramics, shape memory materials (SMM), etc.). Use of ceramic products in industry is constantly growing. Higher hardness, resistance to wear and corrosion, and low coefficient of thermal expansion make these materials essential for certain industries. Conventional machining processes have low efficiency in machining of ceramic products. Although the SEM process can be used only with electrically conductive materials, addition of some conductive components to the structure of ceramic materials makes their machining by SEM possible. Use of modern computer numerical controlled (CNC) machines allows the SEM process to manufacture parts with dimensional accuracy up to 1 μm and positioning accuracy up to 80-100 nm. Computer numerical control over 4-6 independent axes significantly extends the application areas of SEM, whereas modern dielectrics make it safer for the operator's health and the environment. Use of wires or micro-wires as tools to machine micro or nano-sized slots and complicated profiles by WSEM or micro-WSEM processes suffers from issues of frequent wire breakage and streak marks. This leads to process instability, frequent short-circuiting, and, sometimes, damage to the diamond nozzle and the machine

---
*Corresponding author: porvatov_artur@mail.ru

tool. All these factors ultimately result in a machined surface of poor quality (Fuzhu et al. 2007, Ho et al. 2004). This necessitates urgent improvements in SEM and SEM-based processes, through intelligent online monitoring. But the SEM process takes place with the workpiece submerged in the dielectric, and a very small machining zone, making it very difficult and complicated to monitor the CNC machine.

Vibroacoustic (VA) methods have been extensively used for monitoring mechanical machining processes (Gutierrez-Gonzalez et al. 2014, Kozochkin et al. 2014). Herein, the VA signals generated in the machining zone are used. These signals can be used to assess the quality of the machining process, wear, and damage of the cutting tool, in the form of changes in spectral and temporal characteristics (Ravindra et al. 1997). This helps in allowing the machine operator to control the movement of operating elements, timely replacement of the cutting tool, and prevention of accidents and defects. A major advantage of using VA signals is the simplicity of installing accelerometers on the elastic system of the equipment or machine. This enables the user to obtain information from the machining area, even from considerable distance (López-Esteban et al. 2014). Moreover, the problem of placing cables on the machine for transfer of energy and signals can be solved using wireless communication facilities.

There are fifteen main input parameters (Fig. 7.1) used to define the SEM process behavior, which influence the quality of the machined part and the functional aspects of the final product (Kozochkin et al. 2015, Kozochkin and

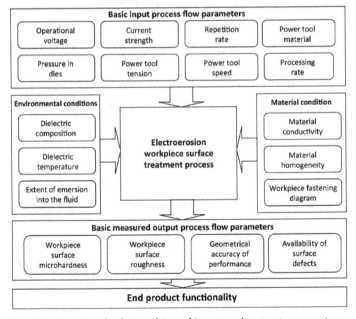

Fig. 7.1: Functional relationships of input and output parameters for SE-based machining processes

Porvatov 2015, and Zhongtao et al. 2015). These parameters can be used to improve the quality of the SEM process by monitoring VA signals. SEM of conductive ceramics has even revealed behavior of VA signals specifically associated with it. Cutting interconnected profiles by SEM along a complex volumetric line in a heavy workpiece (weighing up to 300 kg) requires very strict control of the process to prevent uncontrolled fall of the internal part on the SEM machine and to avoid possible damages to it. Such damages include overcuts, tears, burrs and scratches on the surfaces of SEM machine. Experience reveals that cutting by SEM is accompanied by good quality VA signals as moment of separating the manufactured product from the raw material reaches. Detection and use of these signals allows experienced operators to stop the cutting process in advance to obtain the required separation of the part from the workpiece, without damaging the machine.

7.2 Vibroacoustic Signals in Spark Erosion Based Processes

Figure 7.2 depicts the experimental apparatus for acquiring VA signals in a WSEM process by using two accelerometers (one installed on the working table for workpiece fastening, and the other as an upper guide for the tool electrode). In most cases, however, use of only one accelerometer installed on the worktable is sufficient. Experiments show that the amplitude of VA signals increases gradually as the moment separation of the manufactured product from the raw stock approaches. Subsequently, a series of characteristic peaks

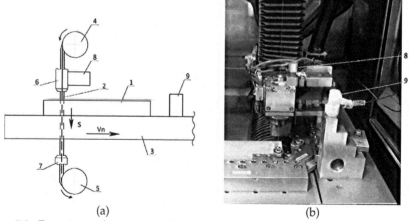

(a)   (b)

Fig. 7.2: Experimental apparatus for acquiring vibroacoustic signals in a WSEM process: (a) schematic diagram, and (b) photograph showing mounting of the accelerometer (1: workpiece; 2: tool electrode; 3: worktable for fixing the workpiece; 4: wire receiving spool; 5: wire giving spool; 6: upper guide; 7: lower guide; 8 and 9: accelerometers; S: direction of winding the wire; Vn: direction of feeding the wire during cutting)

of VA signal is observed during the period of cutting of the workpiece. This indicates weakening of the final linkage of the moving part (under its own mass) and discharge current pulses. At this moment, the part is separated by the tool electrode from the rest of the workpiece, and a series of short circuits occurs. This leads to the formation of burns on the surface of the part. It has also been observed that the weight of the separated part influences the starting moment of part movement and initiation of generation of VA signals. For example, a part weighing 24.5 g (showing mobility) requires fixing with a significant increase in the amplitude of VA signals 5 seconds before the separation of the part, whereas this happens 2 seconds before the separation of the part when it weighs 4.28 g. Mobility of the part is reflected in a wide frequency range of VA signals. However, for low frequencies, VA signals are characterized by instability. Therefore, frequencies higher than 4 kHz are recommended for monitoring the process of part separation or cutting by WSEM.

Figure 7.3 shows the high frequency spectrum of VA signals recorded 30 seconds (spectrum no. 1) and 5 seconds (spectrum no. 2) before part separation in a WSEM process, with the inset showing the octave spectrum of VA signals. Significant increase in amplitude is observed for frequencies higher than 4 kHz. Effective amplitude of the octave-frequency band is more stable because of averaging over a wider frequency range. Effective amplitude at 4 kHz and 8 kHz frequencies increases to twice and one and a half times respectively in the octave-frequency band as the moment of separation approaches. These changes in the amplitude can be used to evaluate the occurrence of an event, and taking decisions regarding the introduction of timely changes in the machining process.

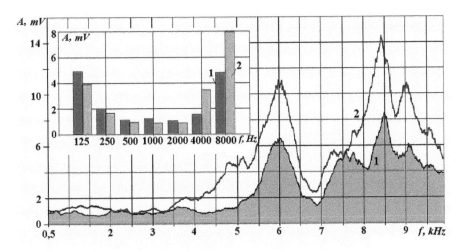

Fig. 7.3: High-frequency spectra of VA signals recorded in a WSEM process, with spectrum no. 1 for 30 seconds before the part separation, and spectrum no. 2 for the last 5 seconds of operation (inset shows octave spectrum)

Intensity of VA signals increases when the symmetry of the position of the tool electrode in the machining gap is upset. This causes prevalence of discharge current pulses on one side of the tool electrode. Such a situation occurs not only during the movement of the part at the moment of its separation, but also when the tool electrode penetrates the workpiece and when the steady state inter-electrode gap (IEG) is not formed around the tool electrode surface. After formation of the steady state IEG, the pulses acting on the tool electrode from its different sides start balancing each other significantly. This decreases the amplitude of VA signals, thus reflecting the stabilization of process parameters. Figure 7.4 depicts two spectra of VA signals obtained during the cutting of an aluminum D16 workpiece by WSEM. Spectrum no. 1 depicts the starting moment of the tool electrode's penetration into the workpiece, and spectrum no. 2 represents the moment just after the formation of the IEG; the inset shows changes in the root mean square (RMS) values of VA signals at the instant of tool electrode penetrating into the workpiece. Similar processes occur during the separation movement of the part under the influence of its own weight and discharge current pulses of the WSEM process. This disturbs the symmetry of the tool electrode's position relative to the sides of the IEG. Subsequently, it increases transverse vibrations of the wire and initiates the appearance of heightened power discharges. Figure 7.5 shows the high frequency spectra of VA signals, with spectrum no. 1 for 30 seconds prior to the end of the process, and spectrum no. 2 for the last second of the process (the inset showing change in the RMS values of the VA signal for the 4 kHz octave band). It can be observed from

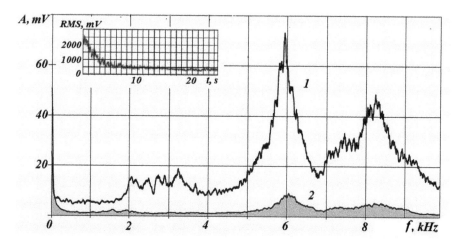

Fig. 7.4: High-frequency spectra of VA signals obtained during cutting of D16 workpiece by WSEM: spectrum no. 1 at the start of the penetration of the tool electrode, and spectrum no. 2 immediately after the formation of a steady-state IEG (inset shows change in the RMS values of VA signals at the moment of tool electrode penetration into the workpiece)

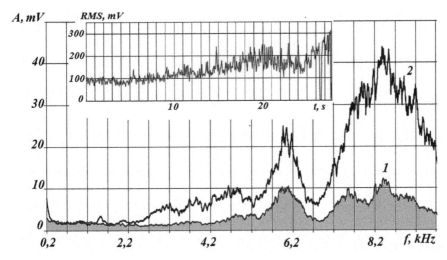

Fig. 7.5: High-frequency spectra of the VA signals obtained at 30 seconds prior to the end of the process (spectrum no. 1) and for the last second of the process (spectrum no. 2) (inset shows the RMS values of the VA signal for the 4 kHz octave band)

Fig. 7.5 that around 8 kHz, the magnitude of spectrum no. 2 (for the last second of the process) increases by a factor of four. Such changes in VA signals are enough for monitoring SEM and WSEM processes, analyzing the collected information, and taking preventive measures to avoid short-circuiting and decrease in machining quality. Excessive amplitudes of VA signals are indicative of an instability in the IEG. This is prorogued by difficulties in removing the products of spark erosion from the IEG, and vibrations of the tool electrode. This tends to decrease the quality of the SEMed or WSEMed surface, and increases burns and wire breaks. Similar information provides the basis for determining the conditions or parameters of SEM or WSEM, especially at the instant when the tool or the wire penetrates the workpiece and at the end of the cut on sides (which decreases disturbing influences on the tool electrode). VA signals stabilize with time during uniform distribution of discharge current pulses, while machining are square, and during uniform removal of spark erosion products.

Decrease in the hardness of the elements of a machining process negatively influences growth of vibrations in it (Pang et al. 2015). Studies on the influence of wire tension on VA signals have revealed that vibrations increase with decrease in wire tension in WSEM, and its influence on the process performance is complicated. It is associated with decrease in the evacuation speed of erosion products from the IEG – with increase in vibrations from one side, and negative influence on surface quality and wear of tool electrode on the other side. Therefore, this requires more intensive research. It has been observed that during stable WSEM using wire tensions of 30 N and 40 N, the RMS values of VA signals disperse with time in both the cases. However, a decrease in wire tension increases the values of these parameters, and such an increase is more noticeable at high frequencies. Influence of process parameters on VA signals is less noticeable when stable machining is disturbed at the onset of the separating movement of the part.

It can be summarized from the aforementioned results and observations that SEM and WSEM processes generate VA signals in a wide frequency range, which can be sensed by the accelerometers mounted on the elastic system of the SEM or WSEM machine (at a distance from the machining zone). VA signals are generated due to (i) disturbances in the tool electrode, caused by discharge current pulses, (ii) cavitation in the machining liquid, and (iii) contact of the tool electrode with the workpiece or erosion products. Monitoring of VA signals allows for reliable tracing of the approach towards the separation moment when a portion or a part is cut from the workpiece, and the instants when the wire or the tool electrode penetrates the workpiece. Spectral contents of VA signals are defined not only by disturbing influences but also by amplitude-frequency characteristics of the elastic elements of the machine tools. A decrease in wire tension or in its hardness increases the amplitude of VA signals in the wide frequency range.

7.3 Monitoring of Spark Erosion Based Processes Using VA Signals

Optimization of productivity, efficiency, surface quality, and dimensional accuracy in SE-based processes demands constant conditions or regulation of process parameters (such as IEG, concentration of erosion products in the IEG, temperature and flow rate of the working fluid). The IEG is a primary parameter as it is largely responsible for determining the quality of a SE-based machining process which occurs within the IEG (Gutkin 1971, Artamonov and Volkov 1991). Even a small increase in the IEG can change the breakdown conditions or even interrupt the discharge, whereas a decrease in the IEG impairs the yield of erosion products, reduces process productivity, promotes build-up of slag on the electrodes, and increases the chances of short-circuiting. SEM is ineffective in the absence of automatic regulation of IEG (Grigor'ev and Kozochkin 2015). While regulating, the IEG must be maintained above the value at which short-circuiting appears, and below the value at which breakdown is impossible and idling pulses appear. The optimal value of IEG is attained when the generation rate of erosion products in the IEG, 'M_p', equals the exit rate of erosion products from the IEG, 'M_{ex}'. M_p is a function of concentration of particles, γ:

$$M_p = f(\gamma) \tag{1}$$

When M_p and M_{ex} are not equal, then the change in concentration is given as:

$$\Delta\gamma = M\frac{\Delta t}{Q} \tag{2}$$

Here, $M = M_p - M_{ex}$, and Q is the volume of the IEG.

For a stable process, $\Delta\gamma = 0$. However, it is very difficult to ensure a constant value of γ, considering the influence of many random factors in SE-based processes. Any fluctuation arising should be promptly eliminated by altering the process parameters by control signals on the basis of diagnostic information. Figure 7.6 shows the variation of generation rate of erosion products 'M_p' with IEG 's' for different exit rates of erosion products 'M_{ex}'. It can be observed from this figure that the maximum value of 'M_p' increases with decrease in 'M_{ex}', due to deterioration in the discharge of erosion products as the tool electrode is introduced in the workpiece and the number of working pulses is reduced (Artamonov and Volkov 1991). Figure 7.7 presents the variation of machining rate M, number of working pulses 'n_w', number of idling pulses 'n_{id}', and number of short-circuit pulses 'n_{sc}', with the percentage of maximum value of IEG, 's_{max}'. It reveals that the IEG value corresponding to the maximum value of machining rate 'M_{max}' is greater than the IEG value corresponding to the maximum value of number of working pulses '$n_{w.max}$'. This is due to an excess of erosion products at '$n_{w.max}$' and the presence of short-circuiting pulses which destabilize the process (Grigor'ev and Kozochkin 2015). Control systems for SE-based machining processes aim to maintain $K_{pu} = 0.7 - 0.9$, where K_{pu} is the efficiency of pulse utilization (calculated as the ratio of the number of working pulses 'n_w' that remove material to the total number of pulses 'n', that is, $K_{pu} = n_w/n$). While K_{pu} is very informative in adaptive control, its use is hindered by the complexity and the inertia of the measuring instruments required. Furthermore, it is difficult to assess efficiency of SE-based machining processes with respect to

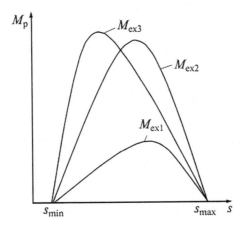

Fig. 7.6: Variation of generation rate of erosion products, 'M_p', with IEG 's' in SE-based processes, for different values of exit rates of erosion products from the IEG, 'M_{ex}' ($M_{ex1} > M_{ex2} > M_{ex3}$)

K_{pu} (Kozochkin et al. 2016, Grigor'ev et al. 2015). When the working fluid is contaminated with erosion products, then the energy of individual pulses is not entirely consumed in the removal of material, for a portion of it is used for the evacuation of erosion products.

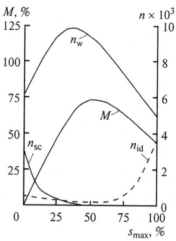

Fig. 7.7: Variation of machining rate 'M', number of working pulses 'n_w', number of idling pulses 'n_{id}', and number of short-circuit pulses 'n_{sc}', with the percentage of maximum value of IEG, 's_{max}'

Some researchers have estimated the number of working pulses 'n_w' from the total width of discharge pulses or from their height at the front and the back of their profiles. This results in an imprecise assessment of the efficiency of SE-based machining processes. A more precise estimate may be obtained by relating efficiency to the ratio of the energy used in the removal of material to the total energy supplied to the machining zone (in the form of discharge pulses). Energy of these discharges is proportional to the effective discharge current 'I_e' for independent generators. This energy may be estimated by means of a current sensor. Parameters of VA signals recorded by an accelerometer mounted on the attachment of the workpiece can be used to assess the energy consumed for material removal. VA signals have been used for monitoring the cutting tool in traditional machining processes (Korenblyum et al. 1977, Nemilov 1989, Kozochkin 2012). However, the comprehensive use of VA signals for monitoring of cutting tool and friction is complicated in a closed system (due to the formation of a nonlinear dependence of VA signals on the contact load), compared to a system having direct contact of the tool and the workpiece, or of the indenter and the counterbody (Schmitz and Smith 2009, Altintas 2000). This nonlinear dependence is not observed when energy pulses act on the workpiece. Hence, a dynamic system more closely resembles a linear model in which dynamic relation between the load source and the workpiece is unchanged. In other words, the model of a dynamic system is significantly simplified and so is its use in the monitoring and regulation of machining process by high-energy fluxes (Grabec and Leskovar 1977, Webster et al. 1996).

Grigor'ev and Kozochkin (2015) obtained VA signals using diode laser impulses and showed that the magnitude of VA signals (and, consequently, the volume of the removed material) and the productivity of an impulse, 'M', increase with an increase in laser power (Figs. 7.8a and 7.8b respectively). It can also be observed from Fig. 7.8b that material removal ceases at very low powers of laser due to the lack of thermal energy required for material sublimation. Increase in productivity 'M' for a laser of moderate power can be approximated by a linear or exponential function with accuracy enough for practical use. The correlation between the amplitude of VA signals and laser impulse productivity can be approximated by the following function:

$$A = C \times M^{\lambda} \tag{3}$$

Here, C and λ are constants.

The results of laser impulses can be used to assume that the parameters of VA signals are related to the SEM process with a similar dependence relationship. Data related to VA signals can be obtained directly from the machining zone, and used to control the SEM process and assess its productivity. The search algorithms for extreme parameters of a machining process do not require any defined mathematical expression which can change under the influence of conditions. One can evaluate the dependence level of 'M' for an SEM process (refer Fig. 7.6) by increasing the size of the IEG by altering the speed of the SEM process and fixing the chosen parameters of VA signals. Gradual adjustment of the IEG along with an evaluation of the amplitude of VA signals allows for an approximation towards its optimal value. If increasing the IEG increases VA signals (at a stable level of current), one can conclude that the optimal value of IEG is more than the current value of IEG. In cases wherein an increase in the IEG decreases VA signals, the approximation is carried out with a gradual increase of the SEM speed. Change in the IEG with respect to the concentration of erosion products can be evaluated by comparing the VA signal level at the present optimal IEG with the preset parameters, and a command can be generated to increase the IEG or the dielectric flushing pressure to remove excess erosion products. The following dependence relation can be used for assessing the productivity of an SEM process, 'M':

$$M = \frac{\beta \cdot w \cdot P}{q \cdot \tau} \tag{4}$$

In the above equation, β is the proportion of usable energy; w is the volume of material removed per unit energy coming with the impulses; P is the energy of the supplied impulses; q is the intermittency of impulses; and τ is the continuance of impulses. However, this relation cannot be considered accurate due to the dependence of 'w' on the energy of the supplied impulses

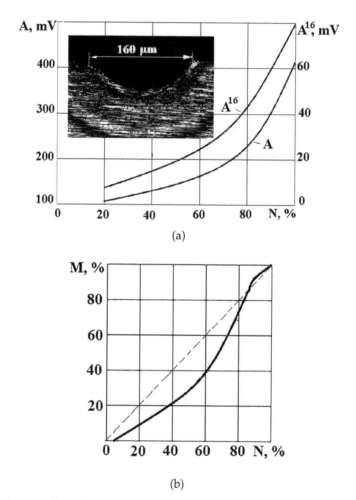

(a)

(b)

Fig. 7.8: Effect of laser power on (a) effective amplitude of VA signals
(A and A^{16} for the 16 kHz octave), and (b) productivity of laser impulse
(inset photo shows the cross-section a drilled well)

(see Fig. 7.8b), and the dependence of 'β' on the concentration of erosion
products, 'γ'. If changes in 'w' are negligible, as in the case of low energy
values, the accidental dependence $\beta = f(\gamma)$ is the main cause for an instability
in SEM.

7.4 Dynamic Modelling of Concentration of Erosion Products

The dynamic model of SE-based processes can be represented as a linear one
with dynamic characteristics dependent on the state of the working fluid.

Influence of the concentration of erosion products on VA signal parameters can be shown by a dynamic model as in Fig. 7.9 and represented by following equation:

$$U(f) = H_1(f) \cdot H_2(f) \cdot Q(f) \tag{5}$$

This model has the following terms:

- A complex sequence of impulses coming from the workpiece-tool combination mounted on the machine of the SE-based process; it is represented as an accidental process $q(t)$, with the amplitude of the signal spectrum as $Q(f)$, f representing frequency
- A frequency response function (FRF) for the working fluid, $H_1(f)$, which changes with the accumulation of erosion products, which in turn varies with the IEG and temperature. The impulses are passed through the operating environment and are modified within it as $q_1(t)$ by the process.
- The dynamic influence of an elastic blank or a workpiece system with dynamic FRF $H_2(f)$

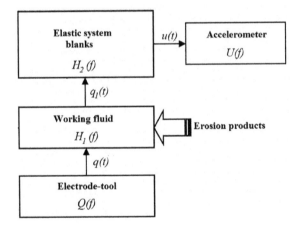

Fig. 7.9: A dynamic model for erosion products in SE-based processes

The accelerometer is mounted on the workpiece to sense vibrations resulting from the abovementioned sources, and has an amplitude $U(f)$, which can be defined according to the linear properties of the dynamic system (Kozochkin 2012, Grigor'ev and Kozochkin 2015). Changes in VA signals depend on the intensity of the energy flow used to do useful work on the surface of the workpiece. A portion of this energy is used for repeated melting and removal of the generated erosion products. Consequently, overlapping craters are created on the surface of the workpiece, causing localization of charges, and heating and vaporization of the workpiece material. The impulses created by phase change of the machined workpiece material are more stretched in time, that is, the concentration of erosion products in the SE-based process is lower for short impulses and higher for longer impulses. Such transformations in the working environment lead to changes in the dynamic characteristic $H_1(f)$, which is included in the FRF ($H(f)$) of the whole channel of observation:

$$H(f) = H_1(f) \cdot H_2(f) \tag{6}$$

One can monitor the values of $H(f)$ in SE-based process by controlling the discharge current and VA signals. Experiments have shown that $H_2(f)$ changes slightly during changes in SE-based processes, but it remains constant during short time periods. Hence, we can consider that all changes in $H(f)$ are related to changes in the working fluid, that is, changes in $H_1(f)$.

7.5 Vibroacoustic Monitoring of Spark Erosion Based Processes

Figure 7.10 presents the FRF for $H(f)$ from data related to VA signals and discharge current during the WSEM of BK60 (M05-ISO) and T15K6 (P10-ISO) alloys for low (Fig. 7.10a) and high (Fig. 7.10b) frequency ranges. Since both input and output signals are measured in the same unit (mV), the FRF values are non-dimensional. Graph 1 corresponds to the initial stage of the WSEM process. During this period, the working fluid is clean, the IEG is located at the surface of the workpiece, and removal of erosion products is relatively facile. The VA signals have the largest amplitudes in the frequency range higher than 11 kHz (Fig 7.10b). Graph 2 shows FRF before the wire breaks, which happens 12 seconds after the start of the WSEM process. Radical changes in the dynamic characteristics of the observed channel are noticeable: reduced amplitude in high-frequency signals and increased amplitudes of the signals in the 2 kHz octave band. These results are consistent with the assumptions about the impacts of discharge localization on the formation of slow discharge pulses. Once the moment of wire-break reaches, a chink is made the workpiece, removal of erosion products becomes more difficult, short impulses decrease, and long impulses appear. FRF changes within the wide frequency range are not convenient for monitoring and control of SE-based processes due to possible interferences from the working equipment. One should only define a few frequency ranges where FRF can be controlled. One can control the effective level of vibration acceleration (RMS value) directly within the most informative frequency range; but it is justified only if the input influence is stable. If the precondition is not fulfilled (control system can change the power of discharging current), it will be more justified to analyze the FRF parameters.

Figure 7.11 presents the RMS value of VA signals in various octave bands from the beginning of the WSEM process till the wire breaks. It reveals a gradual increase in the RMS value of VA signals at low frequencies (graph 1) and a decrease at high frequencies (graph 2). Graph 3 shows the change in the ratio of RMS value at low frequency to that at high frequency, which more clearly demonstrates change of condition during WSEM. It also shows that a stable SEM process is observed until the 7th second, when increase in the RMS value (Graph 3) of VA signals ramps from 3 to 10 times, briefly achieving an increase of 15 times. The control system of machines used for SE-based processes reacts only to indirect electrical parameters, leaving critical increase in the concentration of erosion products out of consideration. This results in overheating of the wire, ultimately leading to its breakage. Use of data only in the form of electrical impulses makes the control system regard the impulses related to erosion products as ineffective, and VA signals from such impulses either cannot be created or have a narrow range.

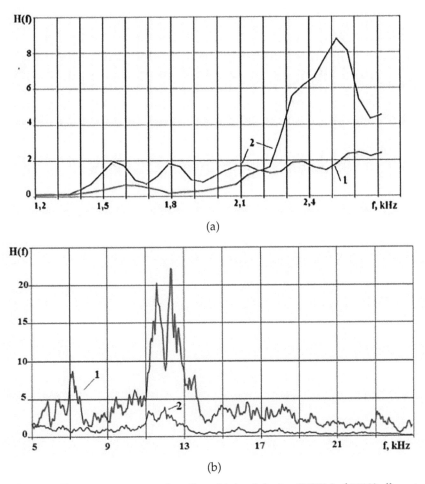

(a)

(b)

Fig. 7.10: Frequency response function obtained during WSEM of BK60 alloy at initial stage (Graph 1) and before the wire breaks (Graph 2) for (a) low frequency range, and (b) high frequency range

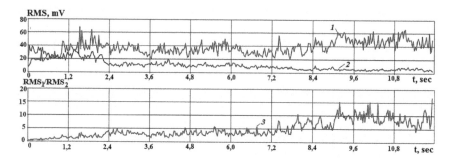

Fig. 7.11: RMS values of VA signals in WSEM of BK60 solid alloy, within octaves of 2 kHz (Graph 1) and 32 kHz (Graph 2), and their interrelation (Graph 3)

Figure 7.12 shows VA signals for 10 ms of WSEM of BK60 alloy, with Graph 1 corresponding to the beginning of the WSEM process, and Graph 2 just before the breakage of the wire. It can be seen from these graphs that the working fluid discharges current impulses in the form of VA signals of less than 0.1 ms duration (Graph 1), but the duration of impulses of VA signals increases up to 5-6 times (Graph 2) at a higher concentration of erosion products. These longer impulses increase the low frequency elements in the spectrum of VA signals.

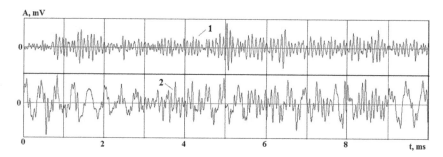

Fig. 7.12: VA signals for 10 ms of WSEM of BK60 alloy: at the beginning of WSEM (Graph 1), and before the breakage of the wire (Graph 2)

These studies show that despite the fact that SE-based processes are non-contact machining processes, they can be monitored using VA signals in the same manner as different contact-type machining processes (Schmitz and Smith 2009, Altintas 2000).

7.6 Vibroacoustic Monitoring of WSEM of Conductive Ceramics

Acquiring and using data related to VA signals for contact-type machining processes is relatively well known (Grabec and Leskovar 1977, Belyi et al. 1981, Webster et al. 1996, Ravindra et al. 1997). Now, it is also being extended for acquiring and using data associated with electric discharge in various devices and for SE-based processes using control of electromagnetic emissions and VA signals (Bhattacharyya and El-Menshawy 1978, Markalous et al. 2008, Smith and Koshy 2013). Further research on the informational features of VA signals during SE-based processing of metals and alloys (Kozochkin et al. 2015, Grigor'ev and Kozochkin 2015, Grigor'ev et al. 2015, Kozochkin et al. 2016) has shown certain specific behaviors of VA signals during the machining of conventional materials, which can be used for monitoring and adaptive control in SE-based processes jointly with electric parameters or even without them. These positive results suggest that the acoustic monitoring of SEM or WSEM of ceramic materials may provide some important information about the state of the SEM or WSEM process, which may prove to be essential for controlling more complex machining procedures for conductive ceramic materials.

This section presents details on the monitoring of WSEM of ceramic, using VA signals. Two types of ceramics have been studied in the this section: *ceramic no. 1* consists of a mixture of aluminum oxide, silicon carbide (in the form of fibers), and conductive titanium carbide, that is, $Al_2O_3+SiC+TiC$, and *ceramic no. 2* comprises a plate made of VOK-60 (K01: group ISO) and a mixture of silicon carbide and conductive titanium carbide, that is, Al_2O_3+TiC. The major difference between these ceramic materials is that more effort is required for achieving maximum homogeneity in the composition of a workpiece made of ceramic no. 1. Comparison of electrical resistance of these ceramic samples with that of hard tool alloys such as T15K6 (P10: group ISO) has revealed that the electrical resistance of 5 mm thick plates made of VOK-60 is 40-50 times higher than that of similar plates made of T15K6. This fact largely defines the specifics of SE-based processing of the ceramic workpieces. Figure 7.13 shows a segment of the recorded root-mean-square (RMS) values of vibration and current signals obtained during the WSEM of ceramic no. 1. A distinct feature of these graphs is that the peaks and valleys of the vibration signal (graph 1) and the current signal (graph 2) are in opposite phase. Some examples have been shown by marked arrows in Fig 7.13.

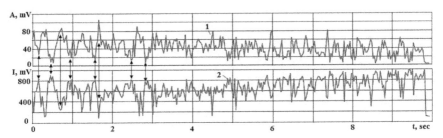

Fig. 7.13: Recorded RMS values of vibration (graph 1) and current (graph 2) signals during the WSEM of ceramic no. 1

Due to the high resistance of the sample of ceramic no.1, WSEM is performed with minimal IEG and is accompanied by frequent short circuits. Figure 7.13 depicts short-circuiting with drop in vibration signals and increase in discharge current signals. The higher resistance of ceramic no.1 ensures moderate voltage and current variation during the short-circuiting, which does not allow the control system of the WSEM machine to identify them in order to take necessary corrective actions. Kozochkin et al. (2016) have shown that increase in the energy and efficiency of pulse action on the WSEMed surface is accompanied with a monotonic growth of the RMS amplitude of high high frequency vibration signals recorded on the workpiece surface. This is confirmed by the fact that a simultaneous drop in VA signal amplitude and growth of current indicates electrode short-circuiting where the electrode temperature increases (with the lack of effective machining). Consequently, in the initial stage of machining, the WSEM electrode (the wire) is overheated and partially burnt-off, subsequently breaking. However, after reducing discharge current pulse frequency and amplitude, ceramic no. 1 is cut, but with a very poor surface quality. Figure 7.14 shows the surface roughness profile of the WSEMed surface of ceramic no. 1, along with its micrograph. It can be observed from this figure that a maximum surface 'R_{max}' value of 67 μm is obtained for an evaluation length of 1.2 mm. The situation is more complicated during the WSEM of ceramic no. 2. Machining is frequently interrupted by short-circuiting, which is accompanied by drops in the RMS amplitude of the VA signal and an increase in the discharge current signal values. The width of the cut for ceramic no. 1 has been found to be 4 mm, while the thickness of the skimmed-off layer came out to be 11 mm for ceramic no. 2. Figure 7.15 presents the signals recorded for vibrations in the 16 kHz octave band and the RMS values of current. A segment of short-circuiting is clearly seen with elevated values of the current signal and minimal values of the vibration (or VA) signal for a duration of 0.35 seconds.

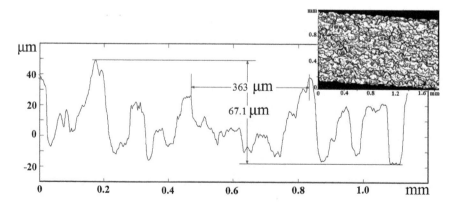

Fig. 7.14: Surface roughness profile of the WSEMed surface of ceramic no. 1 (inset showing its micrograph)

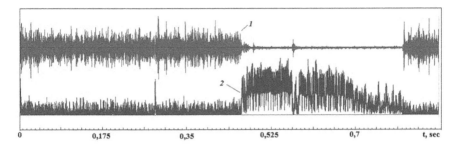

0   0,175   0,35   0,525   0,7   t, sec

Fig. 7.15: Recorded vibration signals (graph 1) and RMS values of current (graph 2) during short-circuiting in the WSEM of ceramic no. 2

Figure 7.16 shows vibration signal spectrums developed during stable WSEM (graph 1) and during short-circuiting of electrodes (graph 2), depicting that frictional contact of the moving electrode with the fixed workpiece is developed during the short-circuit and the accompanying vibration signal spectrum is significantly different than that of stable WSEM. This is true for both low frequency (Fig. 7.16a) and high frequency ranges (Fig. 7.16b). Figure 7.16 also shows that the amplitude of vibrations during stable WSEM is significantly higher than the amplitude during short-circuiting caused by electrode friction. This is particularly noticeable in areas of spectral peaks of vibration or VA signals (areas of natural frequencies of the workpiece and its mounting) associated with WSEM. On appearance of the contact friction, VA signal amplitude is largely defined by the speed of the relative movement of contacting electrode surfaces, and their hardness (Grabec and Leskovar 1977, Kozochkin and Porvatov 2014). Since the rate of wire winding is relatively low (~150 mm/s), the amplitude of vibrations during short-circuiting period is thus smaller than the vibration amplitude occurring during the impact of the discharge current occurring during stable WSEM. However, vibration amplitude during short-circuiting still stands out from the noise produced by the pump and the machine drives. This is an important factor as it allows for the determination of the moment of breaking of contact of electrodes, that is, at the onset of short-circuit, the machining mode control system must instantly send a command to turn-off the discharge pulse generator, turn-on the reverse feed and break the electrode contacts. Only after opening the electrode and the workpiece it is possible to turn-on the pulse generator and restart the feed of the tool electrode. Timely detection of short-circuiting while machining ceramic workpieces is necessary to prevent unacceptable thermal stress in ceramics, and the appearance of brittle cracks.

Due to these issues of identifying short-circuiting in SE-based processing of ceramic materials, the electrodes are in a state of frictional engagement under stress for longer periods of time while machining of ceramic no. 2. Consequently, its WSEM is accompanied with periods of excessive heating of the workpiece surface and formation of brittle cracks.

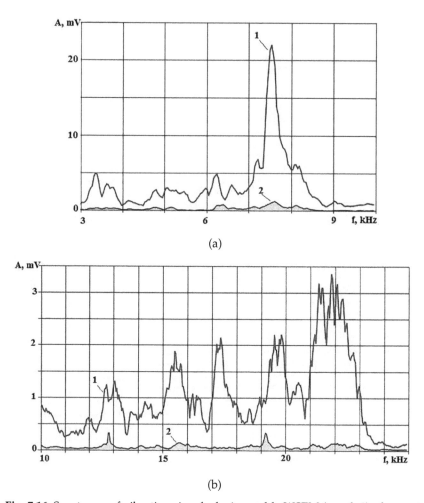

(a)

(b)

Fig. 7.16: Spectrums of vibration signals during stable WSEM (graph 1) of ceramic no. 2 and short-circuiting (graph 2) for (a) low frequency range, and (b) high frequency range

Figure 7.17 shows the photograph of a sample of ceramic no. 2 cut by WSEM. It depicts the formation of cracks and chips on the piece by the wire electrode, near the cut. It can be seen that the formed cracks cause chipping of the ceramic material, with chip size exceeding the width of the cut by several times. During the short-circuit, the wire electrode starts to burn-off, leaving drops of its metallic material on the cut on the ceramic sample. When the electrode contact is broken, these fine filaments of metallic material stretch, maintaining electric contact during the time period when a clearance must be maintained (for geometric reasons), that is, during the withdrawal the electrode from the IEG. Figure 7.18 shows a photograph of a cut obtained during the WSEM of a sample of ceramic no. 2, where solidified drops of metallic material, with fine 5 µm thick wire-like formations, can be seen formed during the process of electrode contact breaking.

Fig. 7.17: Photograph showing the formation of cracks and chips in a sample of ceramic no. 2, by the wire electrode, after its cutting by WSEM process

Fig. 7.18: Photograph of the burnt-off drops of the wire electrode on the cut surface of the workpiece made from sample of ceramic no 2.

Prolonged short-circuits have not been detected during the WSEM of the sample made from ceramic no. 1. However, short-term contact of electrodes occurs frequently. This can be clearly seen from the recorded data as shown in Fig 7.13. Despite the short-term contact of electrodes, the chemical analysis of the sample of ceramic no 1 after cutting by WSEM has shown that, at certain locations, it contained more than 16% of copper. Figure 7.19 shows the location of the wire breaking and the spectrum of the chemical analysis of the surface of sample of ceramic no 2, after the breaking of the wire. Fig. 7.19a clearly displays a fragment of the workpiece material adhering to the wire electrode from the side of the contact of the electrode and the workpiece. Resistance of this adhered material is significantly higher than that of the wire material. Together with the chemical analysis, it can be concluded that the sample contains a layer of the wire material.

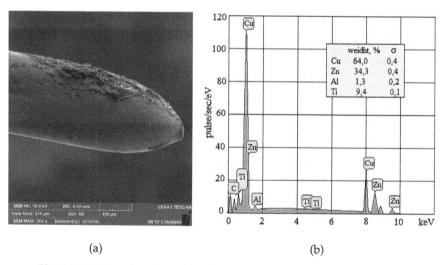

(a)   (b)

Fig. 7.19: (a) Location of wire breaking, and (b) spectrum of the chemical composition of the sample of ceramic no. 2 (after breaking of the wire in the WSEM process)

7.7 Summary

- For wider application of monitoring of SE-based processes using VA signals, it is necessary to conduct more extensive research to get a deep insight into the functional relation of VA signals with machine settings of the SE process, for optimal process parameters, increase in geometric tolerance of a product, etc. It can be also concluded that the use of parameters of VA signals for monitoring of SE-based processes appears rational for improving the quality of solutions of certain problems (for example, operative positioning of the wire electrode in vertical position, the evaluation of the bending of wire during machining in different modes, etc.) that can be extremely important for the manufacturing of parts with complex configurations.
- Study of machining of workpieces with high energy impulses shows that their effectiveness and the parameters of VA signals depend on

the power of impulses supplied, with monotonic dependencies. This allows the monitoring of current effectiveness of SE-based processes and assists in optimizing their parameters. Experiments have shown that parameters of VA signals change due to the dependence of dynamic characteristics of VA signals on the concentration of erosion products in the IEG. Processing of solid alloys with wire cutting has shown that monitoring using VA signals allows for the control of the approach of the breaking moment of the wire.

- SEM or WSEM of conductive ceramic materials is associated with multiple contacts of electrodes, which cannot be identified quickly enough by modern control systems due to certain specific properties of ceramic materials. Prolonged contact of electrodes with a discharge pulse generator turned on leads to cracking of the workpiece, and burning-off and breaking of the tool electrode. It is, however, possible to timely identify moments of electrode short-circuiting by controlling the VA signals. Control of the electrode contact breaking during their separation is also important, contact break can be accompanied with the formation of fine wires from molten metal drops, aggravating the contact breaking process. Such control is also possible using a VA signal analysis. Analysis of VA signals can supplement the existing methods of control of electric parameters, forming a multi-parameter diagnostic system (Kozochkin and Porvatov 2015).

References

Altintas, Y. 2000. Manufacturing Automation: Metal Cutting Mechanics, Machine Tool Vibrations, and CNC Design. Cambridge University Press, Cambridge. UK.

Artamonov, B.A. and Volkov, Y.S. 1991. Analysis of Models of Electrochemical and Electrodischarge Treatment, Part 2: Models of Electrodischarge Treatment, Wire Cutting. Inst Patent Information, Moscow, Russia.

Belyi, V.A., Kholodilov, O.V. and Sviridyonok, A.I. 1981. Acoustic spectrometry used for the evalution of tribological systems. Wear, 69: 309-319.

Bhattacharyya, S.K. and El-Menshawy, M.F. 1978. Monitoring EDM process by radio signals. International Journal of Production Research, 16: 353-363.

Fuzhu, H., Chen, L., Dingwen, Y. and Xiaoguang, Z. 2007. Basic study on pulse generator for micro-SEM. The International Journal of Advanced Manufacturing Technology, 33: 474-479.

Grabec, L. and Leskovar, P. 1977. Acoustic emission of a cutting process. Ultrasonic, 15: 17-20.

Grigor'ev, S.N. and Kozochkin, M.P. 2015. Improvement of machining by the vibroacoustic diagnostics of electrophysical processes. Russian Engineering Research, 35: 801-806.

Grigor'ev, S.N., Kozochkin, M.P., Fedorov, S.V., Porvatov, A.N. and Okun'kova, A.A. 2015. Study of electro-erosion processing by vibroacoustic diagnostic methods. Measurement Techniques, 58: 878-884.

Gutierrez-Gonzalez, C.F., Fernandez-Garcia, E., Fernandez, A., Torrecillas, R. and Lopez-Esteban, S. 2014. Processing, spark plasma sintering, and mechanical behavior of alumina/titanium composites. Journal of Materials Science, 49: 3823-3830.

Gutkin, B.G. 1971. Automated SEM. Mechanical Engineering, Leningrad. USSR.

Ho, K.H., Newman, S.T., Rahimifard, S. and Allen, R.D. 2004. State of the art in wire electrical discharge machining (WEDM). The International Journal of Advanced Manufacturing Technology, 44: 1247-1259.

Korenblyum, M.V., Levit, M.L. and Livshits, A.L. 1977. Adaptive Control of SEM Machines. Nauch. Issled. Inst. Mashinostr, Moscow, Russia.

Kozochkin, M.P. 2012. Nonlinear cutting dynamics. Russian Engineering Research, 32: 387-391.

Kozochkin, M.P., Porvatov, A.N. and Sabirov, F.S. 2014. Vibration testing of technological processes in automated machining equipment. Measurement Techniques, 56: 1414-1420.

Kozochkin, M.P. and Porvatov, A.N. 2014. Effect of adhesion bonds in friction contact on vibroacoustic signal and auto oscillations. Journal of Friction and Wear, 35: 389-395.

Kozochkin, M.P., Grigor'ev, S.N. and Okun'kova, A.A. 2015. Investigation of monitoring perspectives for electroerosion processes by vibration parameter variation. Russian Aeronautics, 58: 488-494.

Kozochkin, M.P. and Porvatov, A.N. 2015. Estimation of uncertainty in solving multi-parameter diagnostic problems. Measurement Techniques, 58: 173-178.

Kozochkin, M.P., Grigor'ev, S.N., Okun'kova, A.A. and Porvatov, A.N. 2016. Monitoring of electric discharge machining by means of acoustic emission. Russian Engineering Research, 36: 244-248.

López-Esteban, S., Bartolomé, J.F., Esteban-Tejeda, L., Moya, J.S., Díaz, L.A., Prado, C., López-Piriz, R. and Torrecillas, R. 2014. Mechanical performance of a biocompatible biocide soda-lime glass-ceramic. Journal of the Mechanical Behavior of Biomedical Materials, 34: 302-312.

Markalous, S.M., Tenbohlen, S. and Feser, K. 2008. Detection and location of partial discharges in power transformers using acoustic and electromagnetic signals. Dielectrics and Electrical Insulation, 15: 1576-1583.

Nemilov, E.F. 1989. Handbook on Electro discharge Treatment of Materials. Mashinostroenie, Leningrad, Russia.

Pang, L., Hosseini, A., Hussein, H.M., Deisab, I. and Kishawy, H.A. 2015. Application of a new thick zone model to the cutting mechanics during end-milling. International Journal of Mechanics Sciences, 96-97: 91-100.

Ravindra, H.V., Srinivasa, Y.G. and Krishnamurthy, R. 1997. Acoustic emission for tool condition monitoring in metal cutting. Wear, 212: 78-84.

Schmitz, T.L. and Smith, K.S. 2009. Machining Dynamics: Frequency Response to Improved Productivity. Springer, New York, USA.

Smith, C. and Koshy, P. 2013. Applications of acoustic mapping in electrical discharge machining. CIRP Annals – Manufacturing Technology, 62: 171-174.

Webster, J., Dong, W.P. and Lindsay, R. 1996. Raw acoustic emission signal analysis of grinding process. CIRP Annals – Manufacturing Technology, 45: 335-340.

Zhongtao, F., Wenyu, Y., Wang, X. and Leopold, J. 2015. Analytical modelling of milling forces for helical end milling based on a predictive machining theory. Procedia CIRP, 31: 258-263.

Improving Sustainability of SE-based Processes

S. Kanmani Subbu[1]* and M.J. Davidson[2]
[1] Department of Mechanical Engineering, IIT Palakkad, Kerala, India
[2] Department of Mechanical Engineering, NIT Warangal, Telangana, India

8.1 Introduction to Sustainability

The Report of World Commission on Environment and Development of United Nations (Brundtland 1987) defines sustainable development as the *"one that meets the needs of the present generations without compromising the ability of future generations to meet their own needs"*. Sustainable development tries to achieve in a balanced manner three pillars, namely, economic development, social development and protection of the environment. The ultimate objective of any manufacturing process is to attain maximum efficiency, effective utilization of energy, reduction of associated costs, and reduction of waste to the environment. The objective of sustainable manufacturing is to optimize process efficiency while simultaneously reducing its negative impacts on the environment and the society (Nambiar 2010). Companies which have adopted sustainable manufacturing concepts are able to produce good quality products, earn more profit and share in the market, and contribute to lower adverse effects on environment and society. Sustainability can be achieved in various stages, such as product stage, system stage and process stage (Deiab 2014). Sustainability at the process stage is achieved by adopting the 6R concept, that is, reduce, reuse, recover, recycle, redesign and remanufacture. This concept is briefly defined as follows (Rajurkar et al. 2017):

- Reduce the use of resources and raw materials
- Reuse the product and its components for the consequent life cycle
- Recover products and components after the first life cycle for reusing them in the subsequent life cycle
- Recycle the scrap generated into raw materials and product

---
*Corresponding author: sksubbu@iitpkd.ac.in

- Redesign the product in such a way that it simplifies the future post-processes
- Remanufacture the used products to re-establish them into products

In cases where machining of certain materials (ceramics, composites, polymers, superalloys) is not possible by conventional machining processes such as turning, milling, shaping, drilling, grinding, etc., non-conventional machining plays a significant role. The latter can machine very hard and/ or brittle, fragile, tough and difficult-to-machine materials by using non-mechanical means such as lower hardness tools, abrasives and/or water jets, electron or laser beams, spark erosion, or electrolytic dissolution. Each non-conventional machining process uses certain properties of the workpiece material; for example, non-conventional machining processes based on electrical energy, and electrolysis-based processes require the workpiece and the tool materials to be electrically conducting. Spark erosion based machining processes such as spark erosion machining (SEM), Wire SEM (WSEM), and processes derived or hybridized from these processes constitute important non-conventional machining processes based on electrical energy, which are used to machine complex shapes out of very hard and/or brittle but electrically conducting materials. Fabrication of complex shaped dies and molds for other manufacturing processes such as casting, forming, and powder metallurgy are their major applications. These processes can also be used for surface alloying, texturing of different engineering components, manufacturing components for electronic industries, aero-engine components, metallic prosthesis, micro-surgical components, and machining of high-alloy materials for turbine applications (Moser 2001). Furthermore, manufacturing of circular and non-circular holes and slots with sharp corners also use this processes (Rajurkar et al. 2017).

8.2 Energy Consumption Aspects in SEM

The SEM process is an electro-thermal process in which material is removed from the workpiece by melting and vaporizing it by means of a precisely controlled series of sparks between the tool and the workpiece, in the presence of an appropriate dielectric fluid. The dielectric fluid helps in concentrating the energy and discharging it efficiently, cooling the tool and the workpiece, solidifying the molten material into particles to constitute the debris, and flushing away the debris from the machining zone (Jain 2009). Since very high temperatures are involved in SEM, it melts the tool material as well and can even decompose or evaporate the dielectric fluid. Consequently, there is always a possibility of generation of some hazardous gases in the SEM process.

Energy required for SEM is more than that for conventional machining processes, at a constant volumetric material removal rate (MRR); for example, SEM requires 150 times more energy than turning for achieving the same value of MRR. This ultimately affects the environment as the consumption of more electrical power (obtained from different types of power plants – hydro, thermal, diesel or nuclear) requires disposal of a larger amount of waste products into the environment. Moreover, the sale of SEM machines is 7% among all machine tools as per the estimates of Charmilles Technologies (Moser 2001), indicating at the 'popularity' of SEM.

Effective energy used in SEM for generating the spark between the tool electrode and the workpiece is approximately 20% of the total energy supplied, whereas the dielectric supply and recirculating system consumes about 50% of the total energy supplied, especially at lower values of peak current (Leao and Pashby 2004). Figure 8.1 shows the contribution of different sources impacting the environment during roughing of hard materials by copper electrode in 1 hour of SEM. It shows that the energy consumption by the SEM process, and the exhaust system contribute 37.4% and 3.1% respectively to the impact on the environment, whereas production and disposal of the dielectric contributes approximately 43.4%. The environmental impact of machine lubrication, generation of compressed gas, process cooling, and removed workpiece is less than 1%, while the removed tool electrode material and the process emissions contribute 8.6% and 5.4% respectively (Kellens et al. 2011). Some researchers (Duflou et al. 2012, Gamage and DeSilva 2015) also confirmed that the dielectric and the energy consumption in SEM impact the environment by 43.4% and 41% respectively.

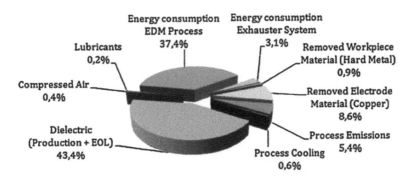

Fig. 8.1: Environmental impact of roughing of hard material by copper electrode in 1 hour of SEM (Kellens et al. 2011)

8.3 Dielectric Fluids in SE-based Processes

Earlier, the dielectric fluid used in SE-based processes included hydrocarbon oils such as kerosene, mineral oil, transformer oil, synthetic oil or silicon oil. Despite their potential advantages in terms of higher productivity, lower tool wear and better surface integrity of the SEMed product, there were several sustainability issues such as by-product gases, vapours, hazardous

smoke, decomposed heavy metals and their products, aerosols, heavy
metallic sharp-edged irregular debris, fire, and exhibition electromagnetic
radiation irrespective of the dielectric used (Rajurkar et al. 2017). Moreover,
SEM is mostly used for different difficult-to-machine metals and alloys such
as nickel, titanium carbide, chromium, tungsten carbide, producing their
micron-sized particles. The size of these particles depends on the combination
of workpiece, tool, and dielectric fluid. Therefore, many protective measures
should be taken for the protection of the health of machine operators, and the
safety of the environment in view of energy consumption.

8.3.1 Desired Characteristics of the Dielectric Fluid

Important properties of the dielectric fluid used in SE-based processes include
viscosity, flash point, pour point, dielectric strength, viscosity, volatility,
evaporation rate, oxidation stability, acid number, colour, odour, effects on
skin, and compatibility (Kern 2009).

- **Flash point:** It is the lowest temperature at which a liquid is ignited in
  the air adjacent to its surface.
- **Pour point:** It is the temperature at which a liquid losses its flow
  characteristics.
- **Dielectric strength:** It is the resistance to breakdown (that is, ionization,
  thereby making a dielectric fluid electrically conducting) of a dielectric
  fluid. Higher dielectric strength of a fluid needs a higher voltage for its
  breakdown.
- **Viscosity:** It is the property of a fluid by virtue of which it resists its flow,
  which is caused by gradual change in shear stresses between its different
  layers.
- **Volatility:** It refers to the tendency of a substance to evaporate at normal
  temperature.
- **Evaporation rate:** A substance of larger area evaporates faster, compared
  to that of a smaller surface area, as it has a higher number of surface
  molecules per unit volume, thus imparting more potential to escape
  from the surface. Moreover, evaporation is faster when the temperature
  of the substance is higher, due to large kinetic energy of the molecules
  present at the surface.
- **Oxidation stability:** It is a type of chemical reaction occurring between
  oxygen and a fluid.
- **Acid Number:** It is a number indicating the acidity of a substance,
  equal to the number of milligrammes of potassium hydroxide needed to
  neutralize the free fatty acids present in one gramme of the substance.

Important aspects to be considered for a sustainable SEM process include
selection of dielectric fluid, its properties with their preferable levels, proper
treatment/disposal of hydrocarbon-based dielectric oils, and alternative
dielectric fluids. Selection of a dielectric also depends on the combination
of the tool and the workpiece material. Preferred values of dielectric

characteristics should be determined for achieving sustainability as well as higher efficiency of the SEM process. The dielectric fluid should have the following characteristics: high flash point, low pour point, high dielectric strength, low volatility, low evaporation rate, high oxidation resistance, low acid number, minimum odour, moderate viscosity, does not produce toxic and harmful gases, does not cause skin irritation, does not react chemically with the components of an SEM machine, and does not emit smoke (Rajurkar et al. 2017).

The recommended concentration limit for total emissions of both aerosol and vapour, and aerosol alone are 20 mg/m^3 and 5 mg/m^3 respectively. Concentration level is measured in different industries and working places by using a sampling device. Maintaining the dielectric level is an important aspect – it should be 40 mm above the erosion spot. Hence, the dielectric can itself absorb the vapour and fumes generated during the SEM process. The effect of carbon, maximum allowable concentration (MAK) value of hydrocarbons, fine dust, formation of carcinogenic by-products as a consequence of catalytic effects, and very high temperatures in the inter-electrode gap (IEG) should be within the permissible limits. Furthermore, the exhausted fumes must not run back into the machining zone without filtering. The sludge formed due to erosion of the slurry which consists of the eroded workpiece and tool materials, solid decomposition products of the dielectric, deionizing resins, and filter cartridges need to be disposed of through suitable methods; otherwise, there is a chance of both soil and water pollution. Wastes from dielectric oils are very poisonous. Therefore, proper environmental regulations have to be followed before the use of a dielectric oil (Rajurkar et al. 2017).

8.3.2 Conventionally Used Dielectric Fluids

Table 8.1 presents the dielectric fluids generally used in SE-based processes, as well as their properties with sustainability issues. Different dielectric fluids emit various substances, fumes and gases during these processes. Some of these are very harmful to the machine operators and to the environment. Table 8.2 presents details of such substances emitted by oil-based and commercial water-based dielectric fluids, and deionized water. Deionized water emits fewer and less hazardous substances than oil-based and water-based dielectric fluids. Oil-based dielectric fluids emit a higher number of and more hazardous substances, especially benzene and benzopyrene (which are carcinogenic), whereas commercial water-based dielectric fluids have intermediate properties in terms of number of substances emitted and their hazardous nature (Leao and Pashby 2004).

8.3.3 Alternative Dielectric Fluids

Alternative dielectric fluids such as water-based dielectric fluids (like tap water, and water mixed with organic compounds), deionized water, gases

Table 8.1: Commonly used dielectric fluids in SE-based processes and their sustainability issues

Dielectric	Properties	Sustainability issues
Kerosene	Low flash point High volatility Low viscosity Good flushing capability	Harmful vapours of CO and CH_4 Skin irritation Odour High risk of fire and explosions
Mineral seal	Petroleum-based product High flash point	Carcinogenic Less expensive Low life span Not recommended
Transformer oil	High flash point High dielectric strength High thermal conductivity High chemical stability	High oxidation rate High rate of sludge accumulation Frequent sludge removal
Deionized water	Odourless High chemical and thermal stability Good flushing capability Environment-friendly	Many cause slight eye and skin irritation
Synthetic oil	High flash point Low volatility and evaporation Long life span	Costly but provides better operator safety and health Less odour

Table 8.2: Substances emitted by different dielectric fluids

Oil-based dielectric fluids	Deionised water	Commercial water-based dielectric fluids
Aldehydes, Acetylene, Butyl acetates, Butyl alcohols, Carbon black Xylene, Carbon monoxide, Carbon dioxide, Ethylene, Hydrogen, Polycyclic aromatic hydrocarbons (e.g., benzopyrene), Paraffinic vapours of hydrocarbons, Oil mists (or aerosol), Metallic particles, and Nitro-aromatic compounds	Carbon monoxide, Chloride, Metallic particles, Nitrogen oxide, Ozone, and Water vapour	Bromide, Benzol, Carbon dioxide, Chlorite, Formaldehyde, Fluorite, Nitrate, Nitrite, Phosphate, Sulphat, Xylene, and Tolune

and non-hydrocarbon-based dielectric fluids, and even dry SEM (using air as the dielectric fluid) are now being used to reduce the impact of conventionally used dielectric fluids on the environment. Alternative dielectric fluids for roughing, finishing and polishing operations should be chosen in such a manner that it results in maximum MRR from the workpiece, and a minimum TWR. The value of MRR is increased due to faster spark disruption with a steeper increase in ignition voltage. In SE-based processes, the workpiece is generally used as the anode. Alternative dielectric fluids do not induce a direct flow of electrons from the cathode to the anode. Therefore, most of the electrons are attracted by the distributed satellite electrodes. Also, they hit the workpiece with lesser kinetic energy and distribute it more than conventional dielectric fluids. These electrodes lead to a faster build-up of the ionization channel, requiring less current to be applied. Thus, a very well-polished workpiece with a surface roughness of less than 0.1 µm can be obtained, which is generally difficult to achieve with conventional dielectric fluids. Moreover, tool wear is reduced by up to 30%, compared to conventional dielectric fluids, due to the effects of ignition time and ignition voltage. The waste particles produced in SE-based processes are expelled explosively from the machining zone in the finest distribution. This reduces short-circuiting and leads to an undisturbed process due to good dispersing qualities of the electric dipoles aligned in the satellite electrodes, causing a faster distribution of waste particles due to their electrical repulsion forces.

Use of plain water as a dielectric results in poor performance in terms of MRR and TWR, compared to the use of hydrocarbon oils. However, results are contradictory when deionized water or even tap water is used as the dielectric fluid, with a brass electrode of negative polarity, and pulse durations smaller than 500 µs (Kern 2009). Use of distilled water as the dielectric fluid in SEM gives better results in terms of MRR and TWR, compared to kerosene. Moreover, use of higher pulse energy results in lower TWR, higher MRR, and good surface finish. Effectiveness of deionized water as a dielectric fluid can be increased by adding organic compounds such as ethylene glycol, polyethylene glycol 600, polyethylene glycol 200, glycerine, polyethylene glycol 400, dextrose, and sucrose. This results in higher MRR and lower TWR than hydrocarbon oils (Kern 2009). Compared to hydrocarbon-based dielectric fluids, emulsion-based dielectric fluids are less expensive, more efficient, and environment-friendly. But these dielectric fluids are less stable than kerosene. Use of an emulsion of water in oil as the dielectric fluid in die-sinking SEM reduces its polluting effects on the environment. Dielectric fluids based on bio-oils and bio-diesels are preferred due to generation of no fumes, environment-friendly nature, and higher efficiency than hydrocarbon-based dielectric fluids (Shaik and Patel 2017). Table 8.3 presents details of some alternative dielectric fluids and their performance conditions.

8.3.4 Important Aspects for Better Handling of Dielectric Fluids

Following are important aspects which must be followed for careful and better handling of dielectric fluids used in SE-based processes in order to

Table 8.3: Performance of some alternative dielectric fluids in SEM

Dielectric medium	Experiment conditions	Machining performance
Distilled water	Workpiece: Ti-6A1-4V Tool: Copper Current: 6 A Polarity: Reverse polarity	Compared to kerosene: • MRR is more • TWR is lower
Distilled water	Workpiece: Low carbon steel Tool: Copper and Brass Pulse duration: 20-500 µs Current: 7.5 A Polarity: Straight and reverse	Compared to kerosene: • Higher MRR with brass as cathode but less with copper as anode. • Surface roughness low with both the electrodes
Distilled water	Workpiece: Steel Tool: Brass Pulse duration: 400-1500 µs	Compared to hydrocarbon oil: • Higher MRR • Lower TWR
Tap water	Workpiece: Low carbon steel Tool: Brass Pulse duration: 20-500 µs Current: 7.5 A Polarity: Straight polarity	Compared to kerosene: • Higher MRR than distilled water or the mixture • Surface roughness is low • No comparison of REW
Air	Workpiece: Steel (S45C) Tool: Copper (ø8.6 mm) Pulse duration: 350 µs Duty factor: 70% at 20A; 280 V	Compared to oil: • No tool wear • Lower MRR
Oxygen mixed with water-based dielectric fluid (Sodick VITOL-QL)	Workpiece: Steel (S45C) Tool: Copper (ø8.6 mm) Pulse duration: 350 µs Duty factor: 70% at 20 A; 280 V	Compared to oil: • No tool wear • Improved MRR
Nitrogen mixed with water-based dielectric fluid (Sodick VITOL-QL)	Workpiece: Steel (S45C) Tool: Copper (ø8.6 mm) Pulse duration: 350 µs Duty factor: 70% at 20 A; 280 V	Not effective
Argon mixed with water-based dielectric fluid (Sodick VITOL-QL)	Workpiece: Steel (S45C) Tool: Copper (ø8.6 mm) Pulse duration: 350 µs Duty factor: 70% at 20 A; 280 V	Not effective

avoid problems pertaining to the safety of operators, and adverse impacts on the environment:

• Proper handling of dielectric fluids begins with careful storage of their packing drums. They always lie down and never stand upright in case they are stored outdoors.

- Equipment such as clean pumps or containers should be used to supply the dielectric fluid to the machine for an SE-based process. Pumps are used for acidic or alkaline dielectric fluids. PVC tubes can also be used sometimes. Anti-corrosive agents, if any, must be removed from the dielectric fluid before supplying it to the machine.
- The machines used for SE-based processes must be cleaned with a few litres of dielectric fluid and should not be cleaned with chlorinated hydrocarbons (Trichloride, Tetrachloroethylene, Trichloroethane, or Freon 12) as the latter form hydrochloric acid. The tool electrode should be cleaned with dry Trichloride before being mounted on the machine.
- If any acid is used to pickle the electrode, then it must not be permitted to get mixed with the dielectric fluid.
- Units for supply and recirculation of dielectric fluid and cooling water should be separate so as to avoid the possibility of water leaking into the dielectric fluid.
- Machines of the SE process should be fitted with a robust hydraulic system to prevent any leakage of hydraulic fluid into the dielectric fluid. The permissible mixing range is 1-2%, beyond which problems occur.
- The machine operator for SE-based processes should have knowledge of the dielectric fluid being used, and should be continuously updated with developments taking place in the area of the dielectric fluid. The finely dispersed electrically conductive particles of the waste material create ionization channels. Hence, fresh dielectric should always be used for at least half an hour before actual machining so that it would no longer fall under the dangerous class of inflammable liquids.

8.4 Flushing and Filtering Aspects in SE-based Processes

Flushing is the process of removing eroded particles from the IEG between the tool electrode and the workpiece, selecting proper flow parameters of the dielectric fluid. If the spark-eroded particles are present between the conducting tool and the workpiece, they form bridges between the tool and the workpiece, causing a short-circuit, and thus creating a big crater on the surface of the tool and the workpiece. Therefore, any further improvement in the performance of an SE-based process necessitates complete removal eroded particles to avoid any chance of short-circuiting. Use of adaptive control in modern machines of SE-based processes enables them to have adaptable pulse interval and variable power supply as per the effectiveness of flushing the IEG.

Dielectric fluid should have lower viscosity and surface tension for better flushing. There are various types of flushing system such as open flushing, pressure or injection flushing, suction or flushed flushing, combined flushing, and interval flushing. Each flushing type has its own advantages and disadvantages. Open or external flushing is simple and is useful when flushing through either the tool or the workpiece is not possible. Pressure or injection flushing applies pressure through a hole in either the tool electrode or the workpiece, for effective removal of debris from the IEG; but the debris in this flushing method moves through the tool or the workpiece and erodes the material, thereby affecting their accuracy. Suction or forced flushing is suitable for producing fine finishing and straight walls on the workpiece. It is used when IEG is very narrow, to ensure that the dielectric is flushed properly to attain a stable process. Very complicated shapes require a combination of pressure and suction type flushing. Interval flushing is applicable for deep hole drilling and fine finishing operations (Jain 2009).

The eroded particles should be removed from the dielectric fluid by cooling it down to normal temperatures and then filtering it properly for high performance of the dielectric fluid. If the temperature of the dielectric is high, then it may evaporate and produce inaccuracies on the workpiece. Filtering is necessary for removing all the particles of debris and dirt, and other foreign particles (like eroded particles from seals, piping, and other machine components) from the used dielectric. Major functions of the filtering unit are to clean the used dielectric coming from the tank, cool the used dielectric, and store and pump the cleaned dielectric into the tank. There are three types of filtering systems – cartridge filtering system, pre-coated filtering system, and transfer filtering system. The filtering effects of cartridge filtering system, pre-coated filter system (with filters), and transfer filter system (without a fine filter) are 1-5 μm, 1 μm, and 1 μm respectively.

8.5 Dry SEM Process

Use of a dielectric fluid in SE-based processes has been proved to increase the performance of the process. The presence of a dielectric concentrates the energy in such processes, thereby enhancing the performance. Moreover, although the vapour bubbles formed in the dielectric liquid tend to expand, the viscosity of the dielectric prevents the expansion, resulting in their explosion; this removes the molten metal from the IEG.

Gases such as argon, helium, compressed air, and oxygen came to be used in SEM processes to overcome the aforementioned problems of a liquid dielectric. The resulting process is referred to as dry spark erosion machining (DSEM). Herein, compressed air or another gas is supplied through a thin-walled tubular pipe to act as a dielectric, to the IEG, thus cooling it and removing the debris generated from sparks.

DSEM reduces the impact on the environment, compared to the environmentally harmful effects associated with liquid dielectric fluids like hydrocarbons and mineral oils. Furthermore, there is no generation of toxic fumes and harmful vapours, or the risk of fire hazard. Other advantages of DSEM include: almost negligible tool wear, low residual stresses, and high precision. Kunieda et al. (1991) and Kunieda and Yoshida (1997) reported that atmospheric pollution reduced drastically when a gas was used as a dielectric, compared to kerosene. Machine tool manufacturers and users are keenly interested in manufacturing simpler machines for SE-based processes by eliminating the complexity of spacious dielectric storage and circulation system for cooling and cleaning of the dielectric fluid. DSEM helps in achieving this by simplifying the manufacturing of machine tools for SE-based processes. It results in reduction of cost and energy required to manufacture machines for SE-based processes (Skrabalak and Kozak 2010). Moreover, the process and the efficiency are also improved in DSEM (Singh and Sharma 2017). These advantages make DSEM to be considered as a green and environment-friendly process compared to SEM. Yeo and New (1999) studied the factors affecting the process performance and the environment, and proposed an analytical method based on process planning approach to choose the process parameters and a suitable dielectric based on utility values. They considered environmental factors along with process parameters to attain the best product quality with less environmental impact.

Gaseous dielectric fluids are mostly used for machining small cavities, resulting in lower MRR and TWR compared to liquid dielectric fluids; moreover, they are independent of pulse duration. Therefore, the use of a gaseous dielectric enables the manufacturing of 3D shapes with high precision. Kunieda et al. (1991) reported that while using oxygen as the dielectric, MRR increases due to an increase in the discharge energy for the oxidation reaction. Yoshida and Kunieda (1999) found that removal and evaporation of the molten material increases and that the solidified material does not get deposited on the workpiece or the tool, and that tool wear is almost zero, regardless of the pulse duration. Yu et al. (2004) observed less deviation on a cemented carbide workpiece and high machining speed in DSEM, compared to when oil was used as the dielectric. Roth et al. (2009) studied MRR and plasma formation for different combinations of tools and workpieces, keeping the parameters of the DSEM process constant. They observed that MRR depends on the workpiece material and the specific energy required for removing the material and for the formation of metallic plasma. Govindan and Joshi (2009) investigated the formation of micro-cracks in DSEM and SEM processes using liquid dielectric fluids. They found that the length and the density of cracks was less in DSEM than in SEM using a liquid dielectric. Beşliu et al. (2010) noted that the accuracy of a circular profile by DSEM process increased when the tool was rotated and translated instead of being translated only. Surface quality also improves when air is used as the dielectric instead of a liquid dielectric. Roth et al. (2013) studied the effects of various gases on the sustainability of the DSEM process and observed that MRR is barely affected by high temperatures due to oxidation and stability and operation efficiency of the process depend on the type of gas used as the dielectric.

8.6 Near-dry SEM Process

Alternatively known as mist-SEM or spray-SEM, near-dry SEM processes use a two-phase mixture of liquid and gaseous dielectric fluids, thus combining the benefits of both. In this type of SEM process, a mist dielectric is supplied to the IEG through a minimum quantity lubrication (MQL) setup (Singh et al. 2016). Spray-SEM uses a dielectric in the form of atomized droplets produced by an ultrasonic atomizer to form a moving thin film that fills the IEG. Hence, there are advantages of droplet-sized dielectric, without compromising with the advantages of a liquid dielectric. Pattabhiraman, et al. (2015) suggested very low consumption of dielectric (5-10 ml/min) for spray-SEM for minimizing energy consumption by the filtration system for cleaning the wastes from the dielectric.

Kao et al. (2007) investigated WSEM and spark erosion drilling (SED) in wet, dry and near-dry conditions, using a mixture of water and air as the dielectric in the near-dry SEM process. Near-dry SEM showed higher MRR, compared to DSEM and SEM (using a liquid dielectric). It also produced sharper cutting edges, less debris deposition, and maintained a small IEG. Tao et al. (2008a) used near-dry SEM for achieving mirror-like finish, using a mist of dielectric delivered through a tubular electrode. They achieved better machining stability and higher surface finish at a lower discharge energy. Surface finish was further improved by reducing the pulse duration. The same group of researchers (Tao et al. 2008b) further studied dry and near-dry spark erosion milling for better MRR in roughing and finishing operations. To achieve a better surface finish, both pulse duration and discharge current should be reduced (Tao et al. 2008b). Gholipoor et al. (2015) studied SED in die steel for various dielectric conditions – dry, near-dry and wet. Their results showed that higher MRR, TWR and surface roughness are achieved using high energy discharge in wet condition (i.e. SED), whereas lower MRR, TWR and surface roughness are achieved using low discharge energy in dry SED (DSED); near-dry SED achieves higher MRR and lower surface roughness, using low discharge energy. The workpiece surface obtained in near-dry SED is better than those obtained from DSED and wet SED.

8.7 Directions for Future Research

Limited research work has been reported on the sustainability aspects of different SE-based processes and the development of related standards for industrial use. Following are some directions for future research towards sustainable manufacturing practices for SE-based processes (Valaki et al. 2015):

- **Selection of the tool-workpiece-dielectric combination:** Selection of dielectric is completely based on the combination of tool and workpiece, for generation of waste and emissions is based on the workpiece-tool-dielectric combination. Therefore, models should be developed in such a way that they forecast the toxicity of generated wastes and characterize the emissions generated during the spark erosion process. The data must provide a choice of the best set of parameters to have the hazardous and toxic emission concentrations within accepted levels in the operator working area.
- **Evaluation of operator risk and mitigation models:** The probability of risks is high in spark erosion process, despite its huge demand in industrial applications. Chances of explosion, and toxic and hazardous working areas primarily affect the safety and the health of the machine operator for any SE-based process. Efficient and systematic investigation is necessary for the assessment of operator risks in terms of possibilities of fire, toxicity level of wastes, and explosions. A model has to be developed in such a way that sets the process parameters to adhere to certain limits to ensure maximum operator safety.
- **Predicting the end of life of dielectric fluid:** Changes in properties of dielectric fluid may occur after its first cycle of use. Use of the same dielectric fluid for subsequent cycles may lead to toxic wastes, hazardous emissions, and fire. Therefore, a scientific method and a suitable model are necessary to recognize the conditions for predicting the end of usable life of reusable dielectric fluids.
- **Reduction in the energy consumption of pumping systems:** Energy consumption in any SE-based process is higher than in other non-conventional machining processes. The pumping system alone consumes up to 70% of energy required by an SE-based process. Therefore, efforts are needed to optimize the components of pumping systems to minimize energy consumption.
- **Comparative assessment of impact on the environment:** Assessment of the impacts of the life cycle of a dielectric on the environment is crucial for all – dry, near-dry and wet – conditions of an SE-based process. This would help in determining the dielectric system with minimum environmental impact and least life-cycle cost. The ISO 14000 family of standards provides practical tools for companies and organizations of all kinds looking to manage their environmental responsibilities. ISO 14649 of this family of standards describes WSEM and defines technology-specific data types representing the machining process of WSEM. The main text of ISO 14649 provides definitions and explanations of the data entities needed to provide control data information to the controller of an SE-based process.

References

Beşliu, I., Schulze, H.P., Coteata, M. and Amarandei, D. 2010. Study on the dry
    electrical discharge machining. International Journal of Material Forming, 3:
    1107-1110.
Brundtland, G.H. 1987. Our Common Future: Report of United Nation's World
    Commission on Environment and Development (UN WCED). Oxford University
    Press, 383 pages (ISBN: 019262080X).
Deiab, I. 2014. On energy efficient and sustainable machining through hybrid
    processes. Materials and Manufacturing Processes, 29: 1338-1345.
Duflou, J.R., Sutherland, J.W., Dornfel, D., Herrmann, C., Jeswiet, J., Kara, S., Hauschild,
    M. and Kellens, K. 2012. Towards energy and resource efficient manufacturing:
    A processes and systems approach. CIRP Annals – Manufacturing Technology,
    61(2): 587-609.
Gamage, J.R. and DeSilva, A.K.M. 2015. Assessment of research needs for sustainability
    of unconventional machining processes. Procedia CIRP, 26: 385-390.
General Assembly Resolution on Implementation of Agenda 21, the Program for the
    Further Implementation of Agenda 21 and the outcomes of the World Summit on
    Sustainable Development (A/RES/64/236).
Gholipoor, A., Baseri, H. and Shabgard, M.R. 2015. Investigation of near dry SEM
    compared with wet and dry SEM processes. Journal of Mechanical Science and
    Technology, 2: 2213-2218.
Govindan, P. and Joshi, S.S. 2009. Analysis of micro-cracks on machined surfaces in
    dry electrical discharge machining. Journal of Manufacturing Processes, 14: 277-
    288.
http://www.SEM-products.com/Dielectrics/ifase/ifase_7.htm
Jain, V.K. 2009. Advanced Machining Processes. Allied Publishers Private Limited,
    India.
Kao, C.C., Tao, J. and Shih, A.J. 2007. Near dry electrical discharge machining.
    International Journal of Machine Tools and Manufacture, 47: 2273-2281.
Kellens, K., Renaldi, Dewulf, D. and Duflou, J.R. 2011. Preliminary environmental
    assessment of electrical discharge machining. Globalized Solutions for
    Sustainability in Manufacturing. pp. 377-382 In: Proceedings of 18th CIRP
    International Conference on Life Cycle Engineering. Hesselbach, J. and
    Herrmann, C. (Eds.). Technical University Braunschweig, Germany, 2-4 May
    2011.
Kern, R. 2009. Sinker dielectric fundamentals. EDM Today (Jan/Feb): 14-20.
Kunieda, M., Furuoya, S. and Aniguchi, N. 1991. Improvement of SEM Efficiency by
    supplying oxygen gas into gap. CIRP Annals – Manufacturing Technology, 40:
    215-218.
Kunieda, M. and Yoshida, M. 1997. Electrical discharge machining in gas. CIRP
    Annals- Manufacturing Technology, 46: 143-146.
Leao, F.N. and Pashby, I.R. 2004. A review on the use of environmentally-friendly
    dielectric fluids in electrical discharge machining. Journal of Materials Process
    Technology, 149: 341-346.
Moser, H. 2001. Growth industries rely on SEM. Manufacturing Engineering, 127:
    62-68.
Nambiar, A.N. 2010. Challenges in sustainable manufacturing. Proceedings of
    the 2010 International Conference on Industrial Engineering and Operations
    Management, Dhaka, Bangladesh.

Pattabhiraman, A., Marla, D. and Kapoor, S.G. 2015. Atomized dielectric spray-based electric discharge machining for sustainable manufacturing. Journal of Micro and Nano-Manufacturing, 3(4): 041008/1-041008/8

Rajurkar, K.P., Hadidi, H., Pariti, J. and Reddy, G.C. 2017. Review of sustainable issues in non-traditional machining processes. Procedia Manufacturing, 7: 714-720.

Roth, R., Balzer, H., Kuster, F. and Wegener, K. 2009. Influence of the anode material on the breakdown behavior in dry electrical discharge machining. Procedia CIRP, 1: 639-644.

Roth, R., Kuster, F. and Wegener, K. 2013. Influence of oxidizing gas on the stability of dry electrical discharge machining process. Procedia CIRP, 6: 338-343.

Shaik, M.B. and Patel, H. 2017. A review on dielectric fluids used for sustainable electro discharge machining. Indian Journal of Scientific Research, 17(2): 40-46.

Singh, J. and Sharma, R.K. 2017. Green EDM Strategies to Minimize Environmental Impact and Improve Process Efficiency. Journal for Manufacturing Science and Production, 16(4): 273-290.

Singh, N.K., Pandey, P.M., Singh, K.K. and Sharma, M.K. 2016. Steps towards green manufacturing through EDM process: A review. Cogent Engineering, 3: 1-8.

Skrabalak, G. and Kozak, J. 2010. Study on dry electrical discharge machining. Proceedings of the World Congress on Engineering. Vol. III. WCE 2010. London.

Tao, J., Shih, A.J. and Ni, J. 2008a. Near-dry SEM milling of mirror-like surface finish. International Journal of Electrical Machining, 13: 29-33.

Tao, J., Shih, A.J. and Ni, J. 2008b. Experimental study of the dry and near-dry electrical discharge milling processes. Journal of Manufacturing Science and Engineering, 130: 011002.

Valaki, J.B., Rathod, P.P. and Khatri, B.C. 2015. Environmental impact, personnel health and operational safety aspects of electric discharge machining: A review. Proceedings of the Institution of Mechanical Engineers, Part B: Journal of Engineering Manufacture, 229(9): 1481-1491.

Yeo, S.H. and New, A.K. 1999. A Method for Green Process Planning in Electric Discharge Machining. A Method for Green Process Planning in Electric Discharge Machining 15: 287-291.

Yoshida, M. and Kunieda, M. 1999. Study on mechanism for minute tool electrode wear in dry SEM. Journal of the Japan Society for Precision Engineering, 65: 689-693.

Yu, Z., Jun, T. and Masanori, K. 2004. Dry electrical discharge machining of cemented carbide. Journal of Materials Processing Technology, 149: 353-357.

Optimization of Parameters of Spark Erosion Based Processes

Rakesh Chaudhari[1], Jay J. Vora[1]*, Alokesh Pramanik[2] and D.M. Parikh[3]

[1] Mechanical Engineering Department, Pandit Deendayal Petroleum University, Gandhinagar, Gujarat, India

[2] Department of Mechanical Engineering, Curtin University, Bentley Campus, Australia

[3] Industrial Engineering Department, Pandit Deendayal Petroleum University, Gandhinagar, Gujarat, India

9.1 Introduction

The ever increasing manufacturing competitiveness makes it essential to develop and adopt new technologies to overcome the challenges of achieving higher accuracy, better quality and surface finish, increased production rate, and enhanced product life with minimum impact on environment, machine operators and society. Furthermore, machining of newly developed exotic smart materials also requires intelligent machining strategies. Since dimensional and geometric accuracy, and surface quality of a machined product primarily depend on the process parameters, it is therefore essential to optimize these parameters (Ghan et al. 2017).

Spark erosion machining (SEM) is a non-conventional machining process which can be used to machine electrically conducting materials, regardless of their mechanical properties, particularly hardness, through a series of sparks generated between the tool electrode and the workpiece, in the presence of a suitably chosen dielectric (Singh et al. 2017). One of the advantages of spark erosion (SE) based processes is that it does not involve direct contact between the tool electrode and the workpiece, thus eliminating the possibility of chatters, mechanical stresses, and vibration problems during material removal (Goyal et al. 2017, Thankachan et al. 2017). Performance of SE-based processes does not depend on the mechanical properties of the workpiece material, especially its hardness and toughness, provided that the workpiece

---

*Corresponding author: vorajaykumar@gmail.com

material is electrically conducting (Pramanik and Littlefair 2016, Mandal et al. 2017, Tan and Yan 2017). Important parameters of such processes, which need to be monitored, controlled and optimized for successful machining, include: discharge or servo-gap voltage (V_s), peak or no load open circuit voltage (V_o), pulse-on time (T_{on}), pulse-off time (T_{off}), discharge current (I), peak current (I_p), type and flushing pressure of the dielectric fluid (P), inter-electrode gap (IEG), and material and polarity of the tool electrode. Additional parameters for WSEM include usability of wire (single-use or reusable), wire material, wire hardness, wire diameter (d), wire tension (T) and wire feed rate (f). Nonetheless, there does exist the possibility of additional parameters specific to a particular SE-based process. Important response parameters or objectives to be optimized (i.e. larger-the-better type maximized or smaller-the-better type minimized) are: material removal rate (MRR), tool wear rate (TWR), dimensional accuracy (i.e. overcut), geometric tolerances or accuracy (i.e. taper), surface roughness (SR) parameters such as average and maximum surface roughness (R_a and R_{max}), and aspects of surface integrity (i.e. microstructure, heat affected zone (HAZ), recast layer thickness, microhardness, residual stresses, wear resistance, and fatigue life) (Satpathy et al. 2017, Senapati et al. 2017). Lockhart and Johnson (2000) defined optimization as *a methodical approach using constraints on the resources and bounds which allow the user to find the optimum or near-optimum solution or set of solutions.* Parametric optimization of an SE-based process generally involves simultaneous optimization of conflicting objectives such as MRR, TWR, SR, dimensional accuracy, geometric accuracy, and surface integrity. For example, increase in current can enhance the MRR, which is desirable, but it also increases the TWR, thus reducing tool life, which is not desirable. In such cases, the optimal solution is the point with a favorable trade-off between MRR and TWR. This is ascertained by developing pareto fronts.

Different optimization methods such as statistical design of experiments approaches (that is, Taguchi method, response surface methodology (RSM)) and meta-heuristics based evolutionary optimization methods have been used to optimize parameters of SE-based processes to enhance their performance. Moreover, optimization methods also help in determining the effects of process parameters on the considered responses of an SE-based process, identifying the most significant parameters, and developing a mathematical model correlating the input and output parameters (Yang et al. 2017).

This chapter presents a review of different optimization approaches used for different SE-based processes. They have been organized into three different groups: (i) the 1st group includes Taguchi design approach, the main focus being the development of system design and parameter design so as to reduce the number of experimental trials for the given set of process parameters; (ii) the 2nd group includes RSM, which gives an optimum set of parameters which can satisfy multiple objectives efficiently, which may not be possible with Taguchi design, and (iii) the 3rd group deals with meta-heuristics based advanced multi-objective optimization approaches which are evolutionary in nature.

9.2 Parametric Optimization of SE-based Processes

Optimization of SE-based process parameters is of great importance, particularly in applications where economy and quality of a machined part is most significant. Appropriate selection of tools, parameters and environmental conditions for a specific feature requires a systematic approach. Different optimization methods used in the published work on optimization of SE-based processes can be classified into the following three categories:

- Design of experiments (DOE) approach such as full factorial DOE, fractional factorial DOE approaches (such as Taguchi method, Box-Behnken design (BBD), central composite design (CCD) of RSM, etc.), leading to the development of statistical models using the experimental results
- Data dependent techniques such as grey relational analysis (GRA), artificial neural network (ANN), fuzzy logic, etc., for the development of intrinsic models using reliable extensive experimental results; these are used when the relationship between the responses and the process parameters is not known at all
- Meta-heuristics based evolutionary optimization methods such as genetic algorithm (GA) and their different versions, simulated annealing (SA), particle swarm optimization (PSO), teaching-learning based optimization (TLBO), artificial bee colony optimization (ABCO), artificial ant colony optimization (AACO), etc.; these methods yield a set of near-optimum solutions rather a single solution

Table 9.1 presents a summary of past research on optimization of parameters of SEM, WSEM, and powder-mixed SEM (PMSEM) processes, using DOE approaches and data dependent techniques, giving details of workpiece material, input parameters, objectives or responses, and optimization method used, along with remarks.

9.2.1 Optimization using Taguchi Design of Experiments Approach

Taguchi's philosophy of quality engineering is the modern offline quality philosophy which involves a three-stage design to determine target values and tolerances of relevant parameters of a process or product (Mitra 1998):

- **System design:** This entails the use of scientific and engineering principles to create a prototype of the product and develop a process that will manufacture/build it.

Table 9.1: Summary of past research on parametric optimization SE-based processes

Researchers and year	Process	Work material	Input variables	Objectives	Optimization method	Remarks
Singh et al. (2004)	SEM	MMC of aluminium alloy reinforced with 10% SiC particles	Pulse-on time Flushing pressure of dielectric Current	MRR TWR SR Taper Radial overcut	Grey relational analysis (GRA)	Best combination of optimum results has the maximum MRR and the minimum value of the rest of the objective parameters.
Tosun et al. (2004)	WSEM	Sodik A320D/EX21	Pulse duration Open circuit voltage Wire-speed Dielectric flushing pressure	MRR Kerf width	Taguchi DOE approach and regression analysis	Pulse duration and open circuit voltage are the most influential parameters for MRR and kerf width.
Hewidy et al. (2005)	WSEM	Inconel 601	Peak current Duty factor Wire tension Dielectric flushing pressure	MRR TWR SR	RSM	MRR increases with increase in peak current and dielectric flushing pressure.
Keskin et al. (2006)	SEM	Steel	Pulse-on time Spark time Power	SR	Taguchi DOE approach	Increase in pulse-on time increases SR.

(Contd.)

Table 9.1: (*Contd.*)

Researchers and year	Process	Work material	Input variables	Objectives	Optimization method	Remarks
Lin et al. (2006)	SEM	SKH 57 high speed steel	Pulse duration Servo-gap voltage Open circuit voltage Peak current Polarity	MRR TWR SR	Taguchi DOE approach	Peak current and polarity influence MRR and TWR, whereas peak current largely affects SR.
Kansal et al. (2006)	PMSEM	AISI D2 die steel	Pulse-on time Pulse-off time Peak current Dielectric flushing pressure Concentration of silicon powder	MRR	Taguchi DOE approach	Peak current and powder concentration are dominant parameters affecting MRR.
Mahapatra and Patnaik (2007)	WSEM	D2 die steel	Pulse duration Pulse frequency Peak current Wire feed rate Wire tension Dielectric flow rate	MRR SR Kerf width	Taguchi DOE approach along with GA	Attempted optimization of non-linear multi-objectives
Çaydaş et al. (2009)	WSEM	AISI D5 tool steel	Pulse duration Open circuit voltage Wire feed rate Dielectric flushing pressure	SR White layer thickness	Adaptive neuro-fuzzy inference system	This approach greatly improves the objective parameters.

El-Taweel (2009)	Die-sinking SEM	Steel	Pulse-on time Current Dielectric flushing pressure Amount of TiC	MRR TWR	RSM	Peak current is the most significant parameter affecting all the objective parameters.
Singh (2012)	SEM	MMC of 6061Al alloy reinforced with 20% Al_2O_3 particles	Pulse-on time Duty cycle Peak current Servo-gap voltage Tool electrode lift time	MRR TWR SR	Taguchi DOE approach and GRA	Pulse current was found to be the most significant amongst all the input parameters.
Gopala-kannan et al. (2012)	SEM	MMC of Al 7075 reinforced with B_4C	Pulse current Pulse-on time Pulse-off time Servo-gap voltage	MRR SR TWR	RSM	Current and pulse-on time significantly influence TWR and SR. TWR decreases with increase in pulse-off time.
Paul et al. (2013)	Micro-SEM	Titanium aluminide alloy	Pulse-on time Pulse frequency Discharge capacitance	Overcut	Taguchi DOE approach	Discharge capacitance significantly affects overcut.
Bobbili et al. (2013)	WSEM	High strength armor steel	Pulse-on time Pulse-off time Servo-gap voltage Wire feed rate Wire tension Dielectric flushing pressure	MRR SR	Taguchi DOE approach	Pulse-on time, pulse-off time, and servo-gap voltage are the important parameters influencing MRR and SR.

(Contd.)

Table 9.1: *(Contd.)*

Researchers and year	Process	Work material	Input variables	Objectives	Optimization method	Remarks
Manjaiah et al. (2014)	WSEM	NiTi shape memory material	Pulse duration Pulse-off time Servo-gap voltage Wire feed rate Dielectric flushing pressure	MRR SR	Taguchi DOE approach	Order of influence of parameters is: pulse duration, dielectric flushing pressure, and pulse-off time.
Tang and Du (2014b)	SEM	Ti6Al4V	Pulse duration Peak current Open circuit voltage Lifting height	TWR MRR SR	Taguchi DOE approach and GRA	Optimized values reduce TWR and SR by 59% and 4%, and increase MRR by 2%.
Kuriachen et al. (2015)	Micro-WSEM	Ti6Al4V	Spark-gap voltage Wire feed rate Wire tension	MRR SR	Fuzzy logic system	Yielded results are reproducible.
Reddy et al. (2015)	SEM	PH17-4 stainless steel	Peak current Surfactant concentration Graphite concentration	MRR SR TWR	Taguchi DOE approach	Order of parameters affecting the objectives is: peak current, surfactant concentration, and powder concentration.

Reference	Technique	Material	Parameters	Responses	Method	Findings
Muthuramalingam and Mohan (2015)	SEM	MMC of Al 6025 reinforced with SiC	Pulse shape, Electrical variables, Discharge energy	MRR, SR, TWR	Taguchi DOE approach	Pulse duration and current are the most influential parameters.
Tiwary et al. (2015)	Micro-SEM	Ti6Al4V	Pulse-on time, Peak current, Servo-gap voltage, Dielectric flushing pressure	MRR, TWR, Overcut, Taper	RSM	RSM model give satisfactory values of all the responses, and its predicted results closely agree with the experimental results.
Aggarwal et al. (2015)	WSEM	Inconel 718	Pulse-on time, Pulse-off time, Servo-gap voltage, Peak current	SR, Cutting rate	RSM	Pulse-on time and servo-gap voltage significantly affect SR, whereas all the parameters except peak current affect cutting rate.
Dewangan et al. (2015)	SEM	AISI P20 tool steel	Pulse-on time, Discharge current	SR, White layer thickness, Surface crack density	GRA and fuzzy logic	The objective parameters are affected by pulse-on time, followed by discharge current.

(Contd.)

Table 9.1: (*Contd.*)

Researchers and year	Process	Work material	Input variables	Objectives	Optimization method	Remarks
Kolli and Kumar (2015)	SEM	Ti-6Al-4V	Discharge current Surfactant concentration Powder concentration	MRR TWR SR Recast layer thickness	Taguchi DOE approach	Current and surfactant concentration significantly affect MRR and TWR, whereas SR and recast layer thickness are affected by discharge current and powder concentration.
Raj and Senthilvelan (2015)	WSEM	Titanium alloy	Pulse-on time Pulse-off time Wire feed rate	MRR SR	RSM	Pulse-on time and pulse-off time largely affect SR, whereas only pulse-off time affects MRR considerably.
Shakeri et al. (2016)	WSEM	Alloy steel	Pulse frequency Peak current Wire feed rate	MRR SR	ANN	ANN is found to be more accurate; GA can also be used for optimization.

Rahang and Patowari (2016)	SEM	Cu-W	Pulse-on time Discharge current Compaction load	TWR MRR SR Edge deviation	Taguchi DOE approach	Used SEM for selective surface modification and indentified optimum parameters for the same
Hsieh et al. (2016)	SEM	Alloys of TiNi and TiNb	Pulse duration Discharge current TiNi composition	MRR TWR SR	RSM	The values of objective parameters increase with pulse duration and current.
Dabade and Karidkar (2016)	WSEM	Inconel 718	Pulse-on time Pulse-off time Servo-gap voltage Discharge current Wire feed rate Wire tension	MRR SR Kerf width Dimensional accuracy	Taguchi DOE approach	Pulse-on time is the most significant parameter affecting all the objective parameters.
Tripathy and Tripathy (2016)	PMSEM	H11 die steel	Pulse-on time Duty cycle Discharge current Servo-gap voltage Powder concentration	MRR TWR TWR SR	TOPSIS and grey relational	Increasing powder concentration up to an appropriate limit leads to a reduction in SR.
Soundararajan et al. (2016)	WSEM	A413 alloy	Pulse-on time Pulse-off time Peak current	MRR SR	RSM	Increase in current and pulse-on time increases MRR, whereas increase in pulse-off time reduces the MRR.

(Contd.)

Table 9.1: (*Contd.*)

Researchers and year	Process	Work material	Input variables	Objectives	Optimization method	Remarks
Dave et al. (2016)	SED	Al 2024 alloy	Pulse-on time Pulse-off time Discharge current	Difference in diameter	RSM and GA	GA combined with RSM helps in successfully optimizing the considered objective parameters.
Shinge et al. (2017)	WSEM	EN31 steel	Pulse-on time Pulse-off time Peak current Wire tension	Machining time SR	Taguchi DOE approach	Peak current considerably affects surface roughness.
Sharma et al. (2017)	WSEM	Ni40Ti60 alloy	Pulse-on time Pulse-off time Servo-gap voltage Discharge current	SR Dimensional accuracy	RSM	Better machining performance in terms of MRR, SR, and dimensional accuracy
Kumar et al. (2018)	WSEM	Titanium	Pulse-on time Pulse-off Time Peak current Servo-gap voltage Wire feed rate Wire tension	MRR Wire wear ratio (WWR)	RSM	Pulse-on and pulse-off times influence WWR, whereas peak current, pulse-on time and pulse-off time influence MRR.

- **Parameter design:** This stage involves the identification of an optimum combination of important controllable parameters of a process/product through fractional factorial design of experiments that minimize variability in its performance by maximizing **signal-to-noise (S/N)** ratio, using the concept of robustness. It is a step procedure in which the *first step* is to choose the parameters of a process/product, which minimize the variability of the objective function (or process/product response), using the concept of robust design. It makes use of non-linear effects of process/product parameters on the responses and creates a design that is robust with respect to the uncontrollable sources of variability. The *second step* is to identify the parameters (referred to as *adjustment parameters*) of the process/product, which have a linear effect on the mean values of responses but do not influence its variability.
- **Tolerance design:** This stage deals with the selection of admissible ranges around target values of the process/parameters identified in the parameter design stage, using the concept of Taguchi loss function. It is done only when performance variation obtained by the combination of optimum values of parameters identified in the parameter design stage is not acceptable.

The term 'signal' as used in *signal-to-noise (S/N) ratio* by Taguchi is the square of the mean value of the desired objective function or process/product response or a quality characteristic; whereas noise is an undesirable term, which is a measure of variability in the considered objective function or process/product response, which is impossible or difficult to control and affects the performance of a product/process. There are three types of noise factors: (i) *Unit-to-unit noise factors:* inherent random variations in the product/process, caused by variability in raw material, machinery, and human resources; (ii) *Internal noise factors:* sources of variation which are internal to a process/product; (a) *time-dependent noise factors* such as wear of mechanical components, spoilage of raw material, and fatigue of parts; (b) *operation related noise factors:* improper settings of the product or machine tools; and (iii) *External noise factors:* sources of variation which are external to the process/product, such as ambient temperature, humidity, wind velocity, input voltage, raw material supply. Internal and external noise factors are referred to as *assignable variations*. Taguchi combined signal and noise terms into one measure, referring to it as *signal-to-noise (S/N) ratio* (the ratio of square of mean of a desirable objective to its standard deviation) for measuring the performance of a process/product. This performance measure has the property that when it is maximized, the expected loss is minimized. Taguchi method tries to select the combination of important controllable process parameters as the optimum solution that will maximize the S/N ratio by minimizing the effects of different types of noise factors using design of experiments (DOE) based on the concept of robustness. Taguchi gave expressions for S/N ratio for three types of objective function:

(i) *Higher-the-better* (MRR, production rate, productivity, profit, consumer satisfaction, etc.), (ii) *Smaller-the-better* (TWR, surface roughness, kerf width, dimensional inaccuracy, geometrical inaccuracy, wear of a component, fuel consumption, environmental pollution, waiting time, etc.), and (iii) *Target-is-best* (production schedule, strength of an engineering component, etc.). Taguchi developed orthogonal arrays as general versions of the Latin square designs. An orthogonal array is a matrix of numbers in which each row represents the level of the chosen parameters of a process/product and each column represents each parameter whose effects on the considered responses of a process/product is to be determined. An orthogonal array has the property that experimental setting of every parameter occurs the same number of times for every experimental setting of all other parameters of the process/product. Moreover, any two columns of an orthogonal array form a two-factor complete factorial design. Use of an orthogonal array minimizes the number of experimental runs, while retaining the pair-wise balancing property at the same time (Mitra 1998, Rosa et al. 2009). Robustness of Taguchi DOE approach and the ease of using its orthogonal arrays have inspired many researchers to successfully use this technique to identify optimum parameters for SE-based processes. The following paragraphs describe some of the research associated with such optimization.

Mahapatra and Patnaik (2007) used Taguchi L_{27} orthogonal array to find the optimum values of six parameters of WSEM process (namely, pulse duration, pulse frequency, peak current, wire feed rate, wire tension, and dielectric flow rate) by varying each of them at three levels to maximize MRR, and minimize SR and kerf width during the WSEM of D2 die steel. They also identified the most critical and mutually interacting parameters for each objective function and reported that the identified optimum combination of the considered parameters is different for each objective function. This is one of the major limitations of Taguchi DOE method. Many researchers have used other data dependent techniques such as GRA, ANN, fuzzy logic, etc., or combined them with Taguchi DOE to overcome the limitations of the Taguchi DOE approach.

Singh et al. (2004) used GRA to identify the optimum values of three parameters of SEM process (that is, pulse-on time, discharge current and flushing pressure of the dielectric) of an MMC of aluminum alloy reinforced with 10% SiC particles, using MRR, TWR, SR, taper, and radial overcut as the objective functions. Their results showed that the best combination of optimum results gave maximum MRR and minimum value of the rest of the objective functions. Kao et al. (2010) used Taguchi L_9 orthogonal array, along with GRA, to optimize four parameters of SEM process (namely, pulse duration, duty cycle, discharge current and open circuit voltage) by varying each of them at three levels and considering MRR, SR and TWR as objective functions during the machining of Ti-6Al-4V alloy. They determined grey relational coefficient and grey relational grades (GRG) from the experimental results, according to GRA method, and reported that the use of GRA significantly simplified the complicated relationship between multiple objectives. Their results showed that discharge current is the most crucial parameter and its smaller value yields lower TWR, SR and MRR. Furthermore, shorter pulse durations reduce surface roughness. The optimized result gave lower TWR and SR, and higher MRR. Results of the confirmation experiments showed decrease in TWR from 0.20 to 0.17 mg/min, increase in MRR from 2.85 to 3.20 mg/min and decrease in SR from 2.20 to 1.78 µm. Kumar et al. (2010) used a combination of Taguchi L_8 orthogonal array and GRA to optimize four parameters of PMSEM process (namely, pulse-on time, duty factor, discharge current, and concentration of silicon abrasive powder) by varying each of them at two levels, maximizing the MRR and minimizing the SR of tool steel. They also reported that GRA simplified the multi-objective optimization by converting it into a single objective function, through GRG, which indicates that the addition of abrasive powder into the dielectric has a more pronounced effect than other process parameters. Results of confirmation experiments showed that the considered objectives greatly improved with the use of GRA. Chen et al. (2010) employed Taguchi L_{18} orthogonal array in conjunction with back-propagation neural network (BPNN) and SA to determine the optimum values of six parameters of WSEM process (that is, pulse-on time, pulse-off time, servo-gap voltage, wire tension, wire feed rate and flushing pressure of the dielectric) by varying each of them at three levels. Experimental results were used to train BPNN for predicting cutting velocity, average surface roughness, and maximum surface roughness valus. They reported this approach to be accurate, on the basis of the claim that error between the prediction and the results of confirmation experiments was less than 3%. Shah et al. (2011) used Taguchi L_{27} orthogonal array to optimize seven parameters of WSEM process (namely, pulse-on time, pulse-off time, servo-gap voltage, material thickness, wire feed rate, wire tension and flushing pressure of the dielectric) by varying each of them at three levels, and using MRR, SR and kerf width as objective functions. Analysis of variance (ANOVA) was done to establish the adequacy of the developed models and to determine the influence of different process parameters on the considered responses. Their results showed that material thickness has negligible effect on MRR, but affects kerf width and SR considerably, and that reduction in spark energy reduces SR and MRR. Rajyalakshmi and Ramaiah (2013) used Taguchi L_{36} orthogonal array along with GRA to optimize the parameters of WSEM process (pulse-on time, pulse-off time, servo-gap voltage, cutting speed, wire tension, wire feed rate and flushing pressure of the dielectric) varying each of them at three levels during the machining of Inconel 825 superalloy. This approach resulted in an increased MRR and reduced SR and spark gap. Also, regression analysis was used to develop mathematical relations between WSEM parameters and the considered responses. Sidhu et al. (2014) used Taguchi L_{27} orthogonal array to investigate the surface modifications of three different types of MMC by PMSEM process and to find the optimum values of pulse-on time, pulse-off time, discharge current, type of dielectric fluid, tool material and workpiece material by varying each of them at three levels. ANOVA of the S/N ratio showed that the micro-hardness of the matrix material was directly proportional to the density of the reinforcing abrasive particles.

Tang and Du (2014a) employed Taguchi L_9 orthogonal array in conjunction with GRA to identify the optimum values of pulse width, discharge current, servo-gap voltage, and lifting height, varying each of them at three levels during the SEM of Ti-6Al-4V alloy. MRR, TWR and SR were considered as objective functions. Their confirmation experiments yielded 87.3% enhancement in MRR, 18.9% decrease in SR and 25.7% decrease in TWR. Kolli and Kumar (2015) used Taguchi L_9 orthogonal array to find the optimum values of discharge current, and concentration of surfactant and graphite powder in the dielectric fluid by varying each of them at three levels during SEM of Ti-6Al-4V alloy, using MRR, SR, TWR and recast layer thickness (RLT) as the objective functions. ANOVA of S/N ratios of the objective functions showed that discharge current and surfactant concentration have significant influence on MRR and TWR, whereas discharge current and graphite powder concentration affect SR and RLT. Lal et al. (2015) also used Taguchi L_{27} orthogonal array combined with GRA to optimize pulse-on time, pulse-off time, discharge current and wire feed rate by varying each of these parameters at three levels in WSEM of an MMC of Al 7075 alloy reinforced with abrasive particles of Al_2O_3 and SiC. They used SR and kerf width as objective functions and concluded that these improved significantly through this approach. ANOVA of the experimental results revealed that all the considered parameters were significant having contribution of 50.02% by pulse-on time, 39.50% by discharge current, 4.58% by pulse-off time, and 2.75% by wire feed rate. Yang et al. (2017) made use of Taguchi L_9 orthogonal array along with GRA to find the optimum values of pulse-on time, pulse-off time, servo-gap voltage and peak voltage by varying each of them at three levels during WSEM of Inconel 718 superalloy, with an objective of maximizing MRR and minimizing SR. They reported that the optimal combination of the parameters derived using GRA resulted in an improvement of 1.8% in SR, with a dual objective function of SR and MRR, compared with the experimental results of L_9 orthogonal array. MRR showed an improvement of 54.5% when Taguchi DOE approach along with GRA was used.

9.2.2 Optimization Using Response Surface Methodology

Response surface methodology (RSM) is a collection of statistical and mathematical tools useful for modeling and analysis of problems in which the response of a process is affected by its controllable parameters and the objective is to optimize it by developing a model for the response surface. If the response can be well modeled by a linear function of controllable parameters, then the developed model is a *first-order model*. If there is a non-linear relationship between response and the controllable parameters, then a polynomial of *second order* or higher is required. RSM is a sequential procedure in which the first step is to decide the limits of the experimental region (or the region of interest) by identifying a range for each parameter. The next step is design and planning of experiments, using a suitable DOE approach of RSM. Central composite design (CCD) is a very efficient and the most commonly used design for fitting a second-order response surface. It consists of 2^k factorial experimental runs: $2k$ axial or star experimental runs at a distance α from the center point of the region of interest, and 3 to 5 central runs where k is the number of process parameters. The second-order model should provide good predictions of the considered response throughout the region of interest and should be rotatable (that is, variance of the predicted response should be constant in a sphere) to achieve that. Rotatability is a reasonable basis for selecting a response surface design as RSM is used for optimization, and location of the optimum or near-optimum solution is not known prior to conducting the experiments. CCD is made rotatable by choosing $\alpha = (n_F)^{1/4}$, where n_F is the number of experimental runs used in the factorial portion of the design. Spherical CCD (SCCD) has $\alpha = \sqrt{k}$, thus putting all factorial and axial points on a spherical surface of radius $\sqrt{k}$. Box-Behnken Design (BBD) is a special case of SCCD, having $\alpha = \sqrt{2}$, and formed by combining 2^k factorial experimental runs with the incomplete block design. It is rotatable or nearly rotatable. These designs are very efficient in terms of the number of required experiments. BBD does not contain any parametric combination in which a parameter is at its lower or upper value in its identified range. One very useful variation of CCD is face-centered cubical design (FCCD), with $\alpha = 1$, in which the experimental region of interest is cuboidal (rather than spherical). This locates the star or axial experimental run on the center of each face of the cube. FCCD requires three levels of each parameter, is not rotatable, and 2 to 3 central points are sufficient to provide good variance of the prediction throughout the experimental region (Montgomery 2004, Box and Draper 1987). Significant amount of work has been carried out using different DOE approaches of RSM for optimizing the parameters of different SE-based processes due to these factors. These studies have been briefly explored in the following paragraphs.

Chiang (2008) used the CCD technique of RSM for experimental investigation and development of models, and optimization of MRR, TWR and OC in terms of four process parameters (pulse-on time, duty factor, open circuit voltage, and discharge current) of SEM of Al_2O_3 and TiC-mixed ceramic. ANOVA was used to investigate the significance of the considered parameters and the interactions between them, using a 95% confidence interval. The results showed that duty factor and discharge current significantly affect MRR, whereas pulse-on time and discharge current affect SR and TWR. Sohani et al. (2009) employed the CCD technique of RSM to study the effects of different shapes of tool electrode (rectangular, triangular, circular and square), to develop models, and optimize MRR and TWR in terms of pulse-on time, pulse-off time, discharge current and tool area in

die-sinking SEM. Results of ANOVA and confirmation experiments reveal
that a circular tool electrode gives higher MRR and lower TWR, followed
by triangular, rectangular and square shapes. Tool area, pulse-off time, and
interaction between pulse-on time and discharge current are statistically
significant for higher MRR and lower TWR. Kung et al. (2009) used the
FCCD technique of RSM to develop models and optimize MRR and TWR in
terms of four parameters (pulse-on time, discharge current, grain size and
concentration of aluminum powder) of PMSEM of cobalt-bonded tungsten
carbide. The results of parametric analysis revealed that MRR increases
with an increase in powder concentration, upto a certain maximum limit,
whereas TWR decreases with decrease in powder concentration. Increase
in grain size, discharge current and pulse-on time increases both MRR and
TWR. Mohanty et al. (2014) used the BBD technique of RSM for experimental
investigation and modelling of MRR and SR during the SEM of Inconel
718 superalloy. The developed models were used for optimization by an
advanced meta-heuristics optimization technique, namely, particle swarm
optimization (Mohanty et al. 2016). The developed models were found to be
robust, as evident from the minimal deviation of the predicted values from
the experimental results. Khundrakpam et al. (2014) used the CCD technique
for experimental investigation and modeling of MRR in terms of discharge
current, silicon powder concentration, and diameter of the tool electrode for
the PMSEM of EN-8 steel. Their results showed that powder concentration
has a major effect on MRR. They concluded that the RSM-developed model
is robust and can predict responses within the specified values of the
experimental domain. Tiwary et al. (2015) employed the CCD technique to
study the influence of pulse-on time, peak current, servo-gap voltage and
flushing pressure of dielectric on MRR, OC, TWR and taper in the micro-
SEM of Ti-6Al-4V alloy and to develop models of the considered responses.
Based on the experimental results, they concluded that the models developed
by RSM predict the responses quite satisfactorily. Sivaraman et al. (2017)
used the CCD technique for experimental investigation, for developing 2nd
order models of MRR and SR (in terms of three parameters, namely, pulse-
on time, pulse-off time and wire tension) and their optimization for WSEM
of titanium alloy. Their results showed that the considered parameters and
some interactions between them have significant effect on SR and MRR.
Results of confirmation experiments revealed that combinations of optimum
parameters show good concurrence with the prediction of response
surface. Sinha et al. (2017) used the BBD technique for the modeling and
optimization of MRR and TWR, in terms of pulse-on time, servo-gap voltage
and discharge current for the SEM of superalloy Incoloy 800, and to study
the effects of the considered parameters on MRR and TWR. Their results
confirmed the robustness of the RSM-based models, due to close agreement
of the predictions with the experimental results.

Similar studies were conducted by Kung and Chiang (2008), El-Taweel
(2009), Habib (2009) and Lin et al. (2012), confirming that RSM is a powerful

modeling technique giving mathematical relationships between parameters and responses of a process, which can be used by meta-heuristics-based evolutionary optimization techniques for both single- and multi-objective optimization.

9.2.3 Optimization Using Evolutionary Optimization Methods

It can be summarized from the abovementioned details of past research on optimization of SE-based processes that an SE-based process has a large number of variable parameters and responses. The relationship between the responses and the process parameters are unknown and non-linear, and the optimization of responses is constrained by limitations of resources and environmental impact. Variable bounds in which the value of a process variable is to be restricted between its feasible minimum (lower) and maximum (upper) values can be considered similar to constraints. This makes the choice of the optimum combination of parameters very challenging, difficult and crucial for optimum or near-optimum performance of the process, and to ensure significant saving of resources and minimize adverse impacts on the environment. The problem is complicated by the fact that most of the responses are conflicting. For example, current increases MRR considerably and, simultaneously, SR and TWR as well, which is not desirable. Taguchi DOE approach optimizes only one response at a time, without considering its effect on the other responses. Whereas the RSM technique gives optimum combination of process parameters without considering the constraints, identifying optimum solution through trial and error method requires a large number of experiments. One approach for solving multi-objective optimization problems involving conflicting objectives is to convert them into a single-objective optimization by assigning weights to each objective function. However, the weighted approach cannot be considered as a global solution as the selection of the weights assigned to the objective function is dependent on designers and application, and is susceptibe to variations. In order to counter this, it has been proposed to probe a set of solutions rather than a single set which suffices the objective function. Also, these sets of solutions are non-dominated by other solutions and can hence be termed as non-dominated solutions. This non-dominated set of solutions is termed as the Pareto front (Konak et al. 2006). These fronts are basically a trade-off between two conflicting objectives. Such problems can be solved by using meta-heuristics-based optimization techniques. In recent years, researchers have developed a number of efficient advanced optimization techniques which have shown their effectiveness in optimizing proces parameters for SE-based processes. Some of these optimization techniques are simulated annealing (SA), genetic algorithm (GA) and its variants for mutli-objective optimization, teaching-learning based optimization (TLBO), particle swarm optimization (PSO), artificial ant colony optimization (AACO), artificial bee colony optimization (ABCO), etc.

Simulated annealing (SA) uses the philosophy of annealing of metals and alloys. It uses a single point search method, unlike other non-traditional optimization techniques. This makes it suffer from drawbacks such as it failure when the objective function becomes discontinuous. *Genetic algorithm (GA)* is based on the genetics philosophy of *'survival of the fittest'*. It uses reproduction, cross-over and mutation operators with their predefined probability to arrive at a set of optimum or near-optimum solutions, starting from a set of solutions (known as population) after the number of predefined generations (or iterations) in a parallel and random manner. GA can be binary-coded or real-coded. Binary-coded GA discretizes the search space, the increment of which depends on the string length used to code a variable parameter which is to be optimized. GA has emerged as a powerful, general-purpose technique widely used for optimization due its several advantages over conventional optimization methods which tend to get trapped at a local optimum value in the search space. It yields an accuracy of more than 90% in predicting process responses as well as optimizing the objective functions. Different variants of GA have been developed to give the Pareto front for multi-objective optimization. They include: non-dominated sorting genetic algorithm (NSGA) and NSGA-II, vector-evaluated genetic algorithm (VEGA) and multi-objective genetic algorithm (MOGA).

Teaching-learning based optimization (TLBO) technique is based on the influence of a teacher on the output of the learners in a class. It is a population-based method and the population is considered as a group of learners. Its working consists of two phases: (i) the 'teacher phase' (learning from a teacher), and (ii) the 'learner phase' (learning from interactions among the learners). A good teacher always trains learners so that they can have improved results, and the learners can also learn through interactions among themselves, group discussions, presentations, etc., which also helps in improving their results. Working of the TLBO algorithm does not require any parameter setting and this is one of its main advantages (Rao et al. 2011).

PSO technique is a global search optimization method, developed by Eberhart and Kennedy (1995), dealing with the behavior of organisms, such as fish schooling and bird flocking. A flock of birds (also called 'particles') represents the solution candidate to an optimization problem. Herein, the position of all the particle with certain velocity needs to be updated in each iteration. The velocity of each particle depends on its best-searched position 'P_{best}' and the position of the best particle in its population 'G_{best}'. All these particles achieve the globally best position after some iterations within the search space. Application of these nature-inspired optimization methods to SE-based process has been carried out by a few researchers. Some noteworthy works in this field have been described in the following paragraphs.

Mandal et al. (2007) used a combination of ANN and NSGA-II for multi-objective optimization by generating Pareto front for MRR and TWR in the SEM of C40 steel (using a copper tool electrode). The generated models were able to predict responses accurately. Rangajanardhaa and Rao (2009) used

ANN and GA in tandem to optimize discharge voltage and peak current, using minimization of SR as the objective function in die-sinking SEM of different workpiece materials (15CDV6, Ti-6Al-4V, M-250 and HE15). GA was used to optimize the weighting factors for ANN. Comparison of experimental and ANN-predicted values of SR revealed errors within acceptable limits. It was also observed that error was reduced from 5% to 2% when ANN was optimized by GA. Tzeng et al. (2011) used a combination of RSM, ANN and GA to optimize the parameters of WSEM of pure tungsten, using an objective function formulated by giving appropriate weights to SR and MRR. Experiments were designed and conducted using RSM, and the experimental results were used to train the ANN for predicting MRR and SR, while GA was used to determine the optimum levels of the considered WSEM process parameters. This approach was found to result in improved prediction as well as confirmation results, compared to those by mere RSM. Amini et al. (2011) used a combination of Taguchi L_{32} orthogonal array, ANN and GA to optimize the parameters of WSEM of TiB_2 nano-composite, using MRR and SR as objective functions. The experiments were conducted using Taguchi L_{32} orthogonal array to study the significance of WSEM process parameters, and the experimental results were used to develop ANN models of MRR and SR. Moreover, GA was used to optimize the weighing functions for ANN models of SR and MRR. Somashekhar et al. (2012) used regression analysis along with SA to optimize the parameters of micro-WSEM, with the objective of maximizing MRR, while minimizing SR and overcut. Regression analysis was used to develop statistical models of MRR, SR and overcut, and, subsequently, a single objective function was formulated by assigning equal weights to them. SA technique was used to identify the optimum values of the considered parameters of micro-WSEM process. Ultimately, the error between optimized and experimental values of response was found to be reasonably small. Mohanty et al. (2016)[Mohanty, 2014 #64] used multi-objective particle swarm optimization (MOPSO) technique to determine the optimum values of pulse duration, duty factor, discharge current, open circuit voltage, electrode material, and flushing pressure of dielectric in SEM, considering MRR, TWR, SR and radial overcut as responses. Six optimization problems were formulated, considering two responses as objective functions and the remaining two responses as constraints. One of the six optimization problems is: Objective Functions: Minimize TWR and Maximize MRR; Subject to: Radial overcut ≤ 0.03; and SR ≤ 6.8, where the minimum values of SR and radial overcut have been extracted from experimental results. These two variables have significant effect on MRR and TWR, provided their magnitudes are restricted by defining the constraints within the specified range. MOPSO suggested various Pareto fronts having non-dominated solutions obtained from the optimization of the six optimization problems formulated. The best solution was selected by composite scores obtained by applying maximum deviation theory (MDT) to the Pareto front optimal solutions. Dang (2017) used Kriging model and MOPSO to maximize MRR and minimize TWR, with a constraint on SR, during the SEM of P20 die steel using a copper tool electrode. Kriging model was used to develop the relationship between SEM process parameters and the responses, and MOPSO was used to generate Pareto front between MRR and TWR. His results reveal a highly non-linear relationship between SEM process parameters and TWR. The generated Pareto front has the optimal parameter settings for achieving the required TWR and MRR.

9.3 Concluding Remarks

This chapter briefly described the commonly used techniques for parametric optimization of SE-based processes, along with a summary of past research on these techniques. We can thus conclude the chapter with the following remarks:

- Taguchi DOE approach has many benefits which make it favorable, particularly for experimental scientists. The main advantage of this technique is the reduction in the required number of experimental trials which incur cost, time and resources for the analysis of the influence of the selected process parameters on the responses. Several input parameters of SE-based processes have been selected or experimented on by different researchers. Taguchi method enables a large number of process parameters to be considered, and their effects on selected responses of the process to be analyzed. Furthermore, the application of ANOVA and GRA to the obtained experimental results helps in quantifying the effect of the selected process parameter on each response. Industrial applications require optimization of multiple objective functions and, in this context, Taguchi method has a major drawback as it provides an individual set of identified optimum parameters for each response; in other words, multi-objective optimization is not possible through Taguchi DOE approach. However, Taguchi method has proved to be effective in providing data for the development of mathematical models by ANN, SA and other multi-objective optimization methods. This approach is particularly useful when a preliminary investigation is to be carried out on some completely new material for which limited or no data/information is available regarding its response to machining.
- RSM approach can be considered a step ahead of Taguchi DOE approach, for it provides mathematical relationships between process parameters and responses by using experimental data. It also reduces the number of experimental trials by using CCD, BBD or FCCD approach. The RSM method investigates the interaction among multiple input parameters and one or more response variables. However, this technique yields a combination of optimum levels of input parameters without considering the constraints.

- Constrained multi-objective optimization problems having conflicting objectives can be solved by meta-heuristics-based and nature-inspired optimization techniques such as SA, GA, TLBO, PSO, etc. by developing Pareto fronts with non-dominated optimum solutions. Pareto fronts, in essence, present a trade-off between two conflicting objectives, and manufacturers can select any point on the front.
- Machining of engineering products and components will always be one of the most important industrial activities. However, in the current scenario, every manufacturing operation is shifting towards sustainability, wherein, apart from the required output, several other aspects such as minimum material wastage, faster machining time, reduced use of energy and reduced impact to society and environment are crucial for manufacturers. Optimization of SE-based process parameters is the need of the hour as it helps in getting maximum output from minimum input. New materials are being developed continuously, with no previous data and information available on their machinability. Machining of these materials using SE-based process is yet to be explored and their process parameters optimized.

References

Aggarwal, V., Khangura, S.S. and Garg, R. 2015. Parametric modeling and optimization for wire electrical discharge machining of Inconel 718 using response surface methodology. The International Journal of Advanced Manufacturing Technology, 79: 31-47.

Amini, H., Soleymani Yazdi, M. and Dehghan, G. 2011. Optimization of process parameters in wire electrical discharge machining of TiB_2 nanocomposite ceramic. Proceedings of the Institution of Mechanical Engineers, Part B: Journal of Engineering Manufacture, 225: 2220-2227.

Bobbili, R., Madhu, V. and Gogia, A. 2013. Effect of wire-EDM machining parameters on surface roughness and material removal rate of high strength armor steel. Materials and Manufacturing Processes, 28: 364-368.

Box, G.E. and Draper, N.R. 1987. Empirical Model-Building and Response Surfaces. Wiley New York.

Çaydaş, U., Hasçalık, A. and Ekici, S. 2009. An adaptive neuro-fuzzy inference system (ANFIS) model for wire-EDM. Expert Systems with Applications, 36: 6135-6139.

Chen, H.C., Lin, J.C., Yang, Y.K. and Tsai, C.H. 2010. Optimization of wire electrical discharge machining for pure tungsten using a neural network integrated simulated annealing approach. Expert Systems with Applications, 37: 7147-7153.

Chiang, K.T. 2008. Modeling and analysis of the effects of machining parameters on the performance characteristics in the EDM process of Al_2O_3+TiC mixed ceramic. The International Journal of Advanced Manufacturing Technology, 37: 523-533.

Dabade, U. and Karidkar, S. 2016. Analysis of response variables in WEDM of Inconel 718 using Taguchi technique. Procedia CIRP, 41: 886-891.

Dang, X.P. 2017. Constrained multi-objective optimization of EDM process parameters using Kriging model and particle swarm algorithm. Materials and Manufacturing Processes, 33(4): 397-404.

Dave, S., Vora, J.J., Thakkar, N., Singh, A., Srivastava, S., Gadhvi, B., Patel, V. and Kumar, A. 2016. Optimization of EDM drilling parameters for Aluminum 2024 alloy using response surface methodology and genetic algorithm. Key Engineering Materials, 706: 3-8.

Dewangan, S., Gangopadhyay, S. and Biswas, C.K. 2015. Multi-response optimization of surface integrity characteristics of EDM process using grey-fuzzy logic-based hybrid approach. Engineering Science and Technology: An International Journal, 18: 361-368.

Eberhart, R. and Kennedy, J.A. 1995. New optimizer using particle swarm theory. Proceedings of IEEE 6th International Symposium on Micro-machine and Human Science 1995 (MHS95), pp. 39-43.

El-Taweel, T. 2009. Multi-response optimization of EDM with Al-Cu-Si-Tic P/M composite electrode. The International Journal of Advanced Manufacturing Technology, 44: 100-113.

Ghan, H.R., Hashmi, A.A. and Dhobe, M.M. 2017. A review on optimization of machining parameters for different materials. International Journal of Advance Research, Ideas and Innovations in Technology, 3(2): 25-27.

Gopalakannan, S., Senthilvelan, T. and Ranganathan, S. 2012. Modeling and optimization of EDM process parameters on machining of Al 7075-B_4C MMC using RSM. Procedia Engineering, 38: 685-690.

Goyal, R., Singh, S. and Kumar, H. 2017. Performance evaluation of cryogenically assisted electric discharge machining (CEDM) process. Materials and Manufacturing Processes, 33(4): 433-443.

Habib, S.S. 2009. Study of the parameters in electrical discharge machining through response surface methodology approach. Applied Mathematical Modelling, 33: 4397-4407.

Hewidy, M., El-Taweel, T. and El-Safty, M. 2005. Modelling the machining parameters of wire electrical discharge machining of Inconel 601 using RSM. Journal of Materials Processing Technology, 169: 328-336.

Hsieh, S.F., Lin, M.H., Chen, S.L., Ou, S.F., Huang, T.S. and Zhou, X.Q. 2016. Surface modification and machining of TiNi/TiNb-based alloys by electrical discharge machining. The International Journal of Advanced Manufacturing Technology, 86: 1475-1485.

Kansal, H., Singh, S. and Kumar, P. 2006. Performance parameters optimization of powder mixed electric discharge machining (PMEDM) through Taguchi's method and utility concept. Indian Journal of Engineering and Materials Sciences, 13(3): 209-216.

Kao, J., Tsao, C., Wang, S. and Hsu, C. 2010. Optimization of the EDM parameters on machining Ti-6Al-4V with multiple quality characteristics. The International Journal of Advanced Manufacturing Technology, 47: 395-402.

Keskin, Y., Halkacı, H.S. and Kizil, M. 2006. An experimental study for determination of the effects of machining parameters on surface roughness in electrical discharge machining (EDM). The International Journal of Advanced Manufacturing Technology, 28: 1118-1121.

Khundrakpam, N.S., Singh, H., Kumar, S. and Brar, G.S. 2014. Investigation and modeling of silicon powder mixed EDM using response surface method. International Journal of Current Engineering and Technology, 4: 1022-1026.

Kolli, M. and Kumar, A. 2015. Effect of dielectric fluid with surfactant and graphite powder on electrical discharge machining of titanium alloy using Taguchi method. Engineering Science and Technology: An International Journal, 18: 524-535.

Konak, A., Coit, D.W. and Smith, A.E. 2006. Multi-objective optimization using genetic algorithms: A tutorial. Reliability Engineering and System Safety, 91: 992-1007.

Kumar, A., Maheshwari, S., Sharma, C. and Beri, N. 2010. A study of multiobjective parametric optimization of silicon abrasive mixed electrical discharge machining of tool steel. Materials and Manufacturing Processes, 25: 1041-1047.

Kumar, A., Kumar, V. and Kumar, J. 2018. Investigation of machining characterization for wire wear ratio and MRR on pure titanium in WEDM process through response surface methodology. Proceedings of the Institution of Mechanical Engineers, Part E: Journal of Process Mechanical Engineering, 232: 108-126.

Kung, K.Y. and Chiang, K.T. 2008. Modeling and analysis of machinability evaluation in the wire electrical discharge machining (WEDM) process of aluminum oxide-based ceramic. Materials and Manufacturing Processes, 23: 241-250.

Kung, K.Y., Horng, J.T. and Chiang, K.T. 2009. Material removal rate and electrode wear ratio study on the powder mixed electrical discharge machining of cobalt-bonded tungsten carbide. The International Journal of Advanced Manufacturing Technology, 40: 95-104.

Kuriachen, B., Somashekhar, K. and Mathew, J. 2015. Multiresponse optimization of micro-wire electrical discharge machining process. The International Journal of Advanced Manufacturing Technology, 76: 91-104.

Lal, S., Kumar, S., Khan, Z.A. and Siddiquee, A.N. 2015. Multi-response optimization of wire electrical discharge machining process parameters for Al 7075/Al_2O_3/SiC hybrid composite using Taguchi-based grey relational analysis. Proceedings of the Institution of Mechanical Engineers, Part B: Journal of Engineering Manufacture, 229: 229-237.

Lin, Y.C., Cheng, C.H., Su, B.L. and Hwang, L.R. 2006. Machining characteristics and optimization of machining parameters of SKH 57 high-speed steel using electrical-discharge machining based on Taguchi method. Materials and Manufacturing Processes, 21: 922-929.

Lin, Y., Tsao, C., Hsu, C., Hung, S. and Wen, D. 2012. Evaluation of the characteristics of the microelectrical discharge machining process using response surface methodology based on the central composite design. The International Journal of Advanced Manufacturing Technology, 62: 1013-1023.

Lockhart, S.D. and Johnson, C.M. 2000. Engineering Design Communication: Conveying Design Through Graphics, Addison-Wesley.

Mahapatra, S. and Patnaik, A. 2007. Optimization of wire electrical discharge machining (WEDM) process parameters using Taguchi method. The International Journal of Advanced Manufacturing Technology, 34: 911-925.

Mandal, D., Pal, S.K. and Saha, P. 2007. Modeling of electrical discharge machining process using back propagation neural network and multi-objective optimization using non-dominating sorting genetic algorithm-II. Journal of Materials Processing Technology, 186: 154-162.

Mandal, A., Dixit, A.R., Chattopadhyaya, S., Paramanik, A., Hloch, S. and Królczyk, G. 2017. Improvement of surface integrity of Nimonic C 263 superalloy produced by WEDM through various post-processing techniques. The International Journal of Advanced Manufacturing Technology, 1-11

Manjaiah, M., Narendranath, S., Basavarajappa, S. and Gaitonde, V. 2014. Wire electric discharge machining characteristics of titanium nickel shape memory alloy. Transactions of Nonferrous Metals Society of China, 24: 3201-3209.

Mitra, A. 1998. Fundamentals of Quality Control and Improvement (2nd Edition). Pearson Education Pte. Ltd. New Delhi.

Mohanty, C.P., Mahapatra, S.S. and Singh, M.R. 2014. An experimental investigation of machinability of Inconel 718 in electrical discharge machining. Procedia Materials Science, 6: 605-611.

Mohanty, C.P., Mahapatra, S.S. and Singh, M.R. 2016. A particle swarm approach for multi-objective optimization of electrical discharge machining process. Journal of Intelligent Manufacturing, 27: 1171-1190.

Montgomery, D.C. 2004. Design and Analysis of Experiments (5th edition), John Wiley & Sons (Asia) Pvt. Ltd. Singapore.

Muthuramalingam, T. and Mohan, B. 2015. A review on influence of electrical process parameters in EDM process. Archives of Civil and Mechanical Engineering, 15: 87-94.

Paul, G., Roy, S., Sarkar, S., Hanumaiah, N. and Mitra, S. 2013. Investigations on influence of process variables on crater dimensions in micro-EDM of Γ-titanium aluminide alloy in dry and oil dielectric media. The International Journal of Advanced Manufacturing Technology, 65: 1009-1017.

Pramanik, A. and Littlefair, G. 2016. Wire EDM mechanism of MMC with the variation of reinforced particle size. Materials and Manufacturing Processes, 31: 1700-1708.

Rahang, M. and Patowari, P.K. 2016. Parametric optimization for selective surface modification in EDM using Taguchi analysis. Materials and Manufacturing Processes, 31: 422-431.

Raj, D.A. and Senthilvelan, T. 2015. Empirical modelling and optimization of process parameters of machining titanium alloy by Wire-EDM using RSM. Materials Today: Proceedings, 2: 1682-1690.

Rajyalakshmi, G. and Ramaiah, P.V. 2013. Multiple process parameter optimization of wire electrical discharge machining on Inconel 825 using Taguchi grey relational analysis. The International Journal of Advanced Manufacturing Technology, 69: 1249-1262.

Rangajanardhaa, G. and Rao, S. 2009. Development of hybrid model and optimization of surface roughness in electric discharge machining using artificial neural networks and genetic algorithm. Journal of Materials Processing Technology, 209: 1512-1520.

Rao, R.V., Savsani, V.J. and Vakharia, D. 2011. Teaching–learning based optimization: A novel method for constrained mechanical design optimization problems. Computer Aided Design, 43: 303-315.

Reddy, V.V., Valli, P.M., Kumar, A. and Reddy, C.S. 2015. Multi-objective optimization of electrical discharge machining of PH17-4 stainless steel with surfactant-mixed and graphite powder-mixed dielectric using Taguchi data envelopment analysis based ranking method. Proceedings of the Institution of Mechanical Engineers, Part B: Journal of Engineering Manufacture, 229: 487-494.

Rosa, J.L., Robin, A., Silva, M., Baldan, C.A. and Peres, M.P. 2009. Electrodeposition of copper on titanium wires: Taguchi experimental design approach. Journal of Materials Processing Technology, 209: 1181-1188.

Satpathy, A., Tripathy, S., Senapati, N.P. and Brahma, M.K. 2017. Optimization of EDM process parameters for AlSiC-20%SiC reinforced metal matrix composite with multi response using TOPSIS. Materials Today: Proceedings, 4: 3043-3052.

Senapati, N.P., Kumar, R., Tripathy, S. and Rout, A. 2017. Multi-objective optimization of EDM process parameters using PCA and TOPSIS method during the machining of Al-20%SiCp metal matrix composite. Innovative Design and Development Practices in Aerospace and Automotive Engineering. Springer.

Shah, A., Mufti, N.A., Rakwal, D. and Bamberg, E. 2011. Material removal rate, kerf, and surface roughness of tungsten carbide machined with wire electrical discharge machining. Journal of Materials Engineering and Performance, 20: 71-76.

Shakeri, S., Ghassemi, A., Hassani, M. and Hajian, A. 2016. Investigation of material removal rate and surface roughness in wire electrical discharge machining process for cementation alloy steel using artificial neural network. The International Journal of Advanced Manufacturing Technology, 82: 549-557.

Sharma, N., Raj, T. and Jangra, K.K. 2017. Parameter optimization and experimental study on wire electrical discharge machining of porous Ni40Ti60 alloy. Proceedings of the Institution of Mechanical Engineers, Part B: Journal of Engineering Manufacture, 231: 956-970.

Shinge, M., Mithari, R., Rajigare, P. and Patil, K. 2017. Optimization of process parameters of wire electrical discharge machining of EN31 steel. Pulse, 1: 3.

Sidhu, S.S., Batish, A. and Kumar, S. 2014. Study of surface properties in particulate-reinforced metal matrix composites (MMC) using powder-mixed electrical discharge machining (EDM). Materials and Manufacturing Processes, 29: 46-52.

Singh, M.A., Sarma, D.K., Hanzel, O., Sedláček, J. and Šajgalík, P. 2017. Machinability analysis of multi walled carbon nanotubes filled alumina composites in wire electrical discharge machining process. Journal of the European Ceramic Society, 37: 3107-3114.

Singh, P.N., Raghukandan, K. and Pai, B. 2004. Optimization by grey relational analysis of EDM parameters in machining Al-10%SiCp composites. Journal of Materials Processing Technology, 155: 1658-1661.

Singh, S. 2012. Optimization of machining characteristics in electric discharge machining of 6061Al/Al$_2$O$_3$P/20p composites by grey relational analysis. The International Journal of Advanced Manufacturing Technology, 63: 1191-1202.

Sinha, S., Ballav, R. and Kumar, A. 2017. Investigation on material removal rate and tool wear rate in electrical discharge machining of Incoloy 800 by using response surface methodology. Materials Today: Proceedings, 4: 10603-10606.

Sivaraman, B., Padmavathy, S., Jothiprakash, P. and Keerthivasan, T. 2017. Multi-response optimisation of cutting parameters of wire EDM in titanium using response surface methodology. Applied Mechanics and Materials, 854: 93-100.

Sohani, M., Gaitonde, V., Siddeswarappa, B. and Deshpande, A. 2009. Investigations on the effect of tool shapes with size factor consideration in sink electrical discharge machining (EDM) process. The International Journal of Advanced Manufacturing Technology, 45: 1131-1145.

Somashekhar, K., Mathew, J. and Ramachandran, N. 2012. A feasibility approach by simulated annealing on optimization of micro-wire electric discharge machining parameters. The International Journal of Advanced Manufacturing Technology, 61: 1209-1213.

Soundararajan, R., Ramesh, A., Mohanraj, N. and Parthasarathi, N. 2016. An investigation on material removal rate and surface roughness of squeeze casted A413 alloy in WEDM by multi-response optimization using RSM. Journal of Alloys and Compounds, 685: 533-545.

Tan, T.H. and Yan, J. 2017. Atomic-scale characterization of subsurface damage and structural changes of single-crystal silicon carbide subjected to electrical discharge machining. Acta Materialia, 123: 362-372.

Tang, L. and Du, Y. 2014a. Experimental study on green electrical discharge machining in tap water of Ti-6al-4V and parameters optimization. The International Journal of Advanced Manufacturing Technology, 70: 469-475.

Tang, L. and Du, Y. 2014b. Multi-objective optimization of green electrical discharge machining Ti-6Al-4V in tap water via grey-Taguchi method. Materials and Manufacturing Processes, 29: 507-513.

Thankachan, T., Prakash, K.S. and Loganathan, M. 2017. WEDM process parameter optimization of FSPED copper-Bn composites. Materials and Manufacturing Processes.

Tiwary, A., Pradhan, B. and Bhattacharyya, B. 2015. Study in the influence of micro-EDM process parameters during machining of Ti-6Al-4V superalloy. The International Journal of Advanced Manufacturing Technology, 76: 151-160.

Tosun, N., Cogun, C. and Tosun, G. 2004. A study on kerf and material removal rate in wire electrical discharge machining based on Taguchi method. Journal of Materials Processing Technology, 152: 316-322.

Tripathy, S. and Tripathy, D. 2016. Multi-attribute optimization of machining process parameters in powder mixed electro-discharge machining using TOPSIS and grey relational analysis. Engineering Science and Technology: An International Journal, 19: 62-70.

Tzeng, C., Yang, Y., Hsieh, M. and Jeng, M. 2011. Optimization of wire electrical discharge machining of pure tungsten using neural network and response surface methodology. Proceedings of the Institution of Mechanical Engineers, Part B: Journal of Engineering Manufacture, 225: 841-852.

Ubaid, A.M., Dweiri, F.T., Aghdeab, S.H. and Al-Juboori, L.A. 2017. Optimization of EDM process parameters with fuzzy logic for stainless steel 304 (ASTM A240).

Yang, C.B., Lin, C.G., Chiang, H.L. and Chen, C.C. 2017. Single and multi-objective optimization of Inconel 718 nickel-based superalloy in wire electrical discharge machining. The International Journal of Advanced Manufacturing Technology, 1-10.

Index